Charles Theodore Williams

The Climate of the South of France

Second Edition

Charles Theodore Williams

The Climate of the South of France
Second Edition

ISBN/EAN: 9783337254001

Printed in Europe, USA, Canada, Australia, Japan

Cover: Foto ©Andreas Hilbeck / pixelio.de

More available books at **www.hansebooks.com**

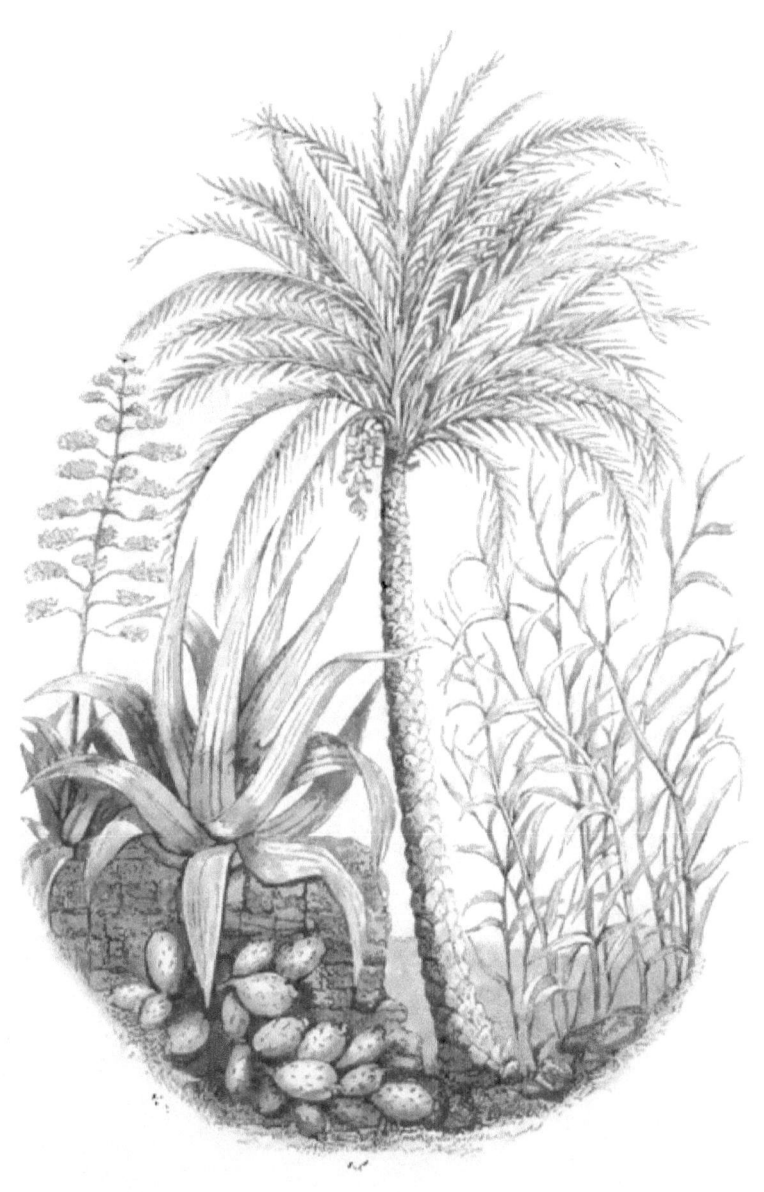

TROPICAL VEGETATION OF THE SOUTH OF FRANCE.

THE CLIMATE

OF THE

SOUTH OF FRANCE

AS

SUITED TO INVALIDS:

WITH NOTICES OF

MEDITERRANEAN AND OTHER WINTER STATIONS.

BY

CHARLES THEODORE WILLIAMS,

M.A., M.D. OXON.

ASSISTANT-PHYSICIAN TO THE HOSPITAL FOR CONSUMPTION AND
DISEASES OF THE CHEST AT BROMPTON.

SECOND EDITION.

LONDON:
LONGMANS, GREEN, AND CO.
1869.

THE CLIMATE

OF THE

SOUTH OF FRANCE

AS

SUITED TO INVALIDS:

WITH NOTICES OF

MEDITERRANEAN AND OTHER WINTER STATIONS.

BY

CHARLES THEODORE WILLIAMS,

M.A., M.D. OXON.

SENIOR ASSISTANT-PHYSICIAN TO THE HOSPITAL FOR CONSUMPTION
AND DISEASES OF THE CHEST AT BROMPTON.

SECOND EDITION,

With an Appendix

ON

ALPINE SUMMER QUARTERS FOR INVALIDS,
AND ON THE MOUNTAIN CURE.

LONDON:
LONGMANS, GREEN, AND CO.
1870.

LONDON: PRINTED BY
SPOTTISWOODE AND CO., NEW-STREET SQUARE
AND PARLIAMENT STREET

PREFACE.

The object of the first edition of this book was to give a brief and impartial survey of the Climate of the South of France, and of its varieties best suited to pulmonary invalids; and the fact that a second edition is now called for, shows that such an account, when written in a spirit free from local interest or bias, has proved acceptable to a large number of readers.

My information has been derived partly from friends resident in the localities described, but chiefly from my own personal observations made during four visits to, and a prolonged sojourn in, this beautiful region.

To render the little work more complete I have enlarged its boundaries, and it now contains descriptions of the principal Mediterranean winter stations, including those of France, Italy, Spain, and North Africa. Notices of some summer stations have been inserted in an Appendix; and, to render the whole subject more intelligible, a map has been added.

A chapter has been devoted to the hygienics of consumption, in which will be found practical hints as to the invalid's life in the south of Europe, and his food, clothing, exercise, &c.

I have to acknowledge valuable aid from the experience and suggestions of my father, Dr. C. J. B. Williams; and my brother, Mr. H. S. Williams, late Scholar of St. John's College, Cambridge, has greatly assisted me in the meteorological portion of the work. In connection with this subject I think it right to mention, that although the general results of meteorological observations alone have been stated, the ample data which supply them have been carefully examined, but are too voluminous for publication.

78 PARK STREET,
    GROSVENOR SQUARE:
       September 1869.

CONTENTS.

CHAPTER I.

GENERAL ADVANTAGES OF THE SOUTH OF FRANCE.

PAGES

Accessibility—Sunshine and fine weather—Facilities for outdoor exercise—Influence on the mind—Medical staff—Disadvantages of the climate—Drainage—Aspects of climate divided into Physical and Medical—Dry earth system 1–10

CHAPTER II.

PHYSICAL ASPECTS.

Thermometrical phenomena—Comparison of Nice with Torquay and the Cove of Cork—Hyères and Kew—Sources of warmth—The sun—Day and night temperature—Mediterranean Sea—Absence of tide—Its saltness—Cause—Its warmth—Causes—Soundings of Admiral Smyth, Captain Spratt, and others—Contrast with Atlantic—Influence of seasons on temperature—Effect on vegetation. *Hygrometrical phenomena* — Rainfall and rainy days—Atmospheric moisture — Hygrometrical comparison of Mentone, Hyères, Nice, and Kew. *Other meteorological phenomena*—Winds — Mistral—Force — Causes — Theories of M. Martins and Dr. Bennet—Probable cosmical origin of the mistral—Southerly and other winds—Influence of neighbouring mountain ranges 11–37

CHAPTER III.

HEALTH RESORTS OF THE REGION.

HYÈRES—Distance from sea—Luxuriant vegetation—Palms—Wild flowers—Iles d'Hyères—Military hospital—Hotels—Its less dry and exciting climate—Costabelle—Its high mean temperature. CANNES—Its beautiful site—Proximity to the sea—Vegetation—Scent-producing plants—Dry and stimulating climate—Hotels—Cannet—Its inland and softer climate. NICE—Its situation and shelter—Its suburbs, Cimiez, Carabaçel, and St. Barthélemy—Their more sheltered position, and vegetation—Dry climate of Nice—Its vicissitudes—Its beneficial effect in some diseases—Intermittent fever—Its origin—Moister climate of Carabaçel and Cimiez—Its resemblance to that of Hyères—Hotels—Monaco. MENTONE—Rich vegetation—Lemons—Carob trees—Complete shelter from northerly winds—Climate warmest and driest of the health resorts—Its defects—Proximity to sea—Closeness of atmosphere—The two Bays—Inland accommodation—Facilities for outdoor exercises in the different health resorts . . 38–61

CHAPTER IV.

MEDICAL ASPECTS—EFFECTS OF THE CLIMATE ON HEALTH AND DISEASE.

Acute disease common—Chronic and degenerative rare—M. Richelmi's experience—Low death-rate of Consumption—Dr. Chambers's Genoa and London statistics—Negative effects of the climate—Positive effects—Stimulating influence—Causes—Good and bad results—Choice of a health resort—Bronchitis, humid and dry—Asthma—Phthisis—Mediterranean climate, where beneficial, where hurtful—British watering-places. PAU—Its calm atmosphere—Its climate compared with that of Nice. MADEIRA—Decreased popularity—

PAGES

Causes, alleged and real—Prevalence of fever—Introduction of the sugar-cane—Soft nature of climate—Experiment of Brompton Consumption Hospital . . 62–75

CHAPTER V.

HYGIENICS OF CONSUMPTION IN THE SOUTH OF FRANCE.

Food, &c.—Cooking—Cod-liver oil—Vegetables—Their abundance and excellence—Fruit—Alcoholic drinks—Tonics. *Clothing*—Importance of warm clothes. *Ventilation*—Pure air—Temperature to be considered—Sleeping with open windows—Its danger. *Exercise*—Its benefits—Active exercise—Its varieties—Rowing—Swinging—Climbing—Gymnastics—Walking—Mountain ascents—Passive exercise—Carriage—Sailing—Their effects—Riding 76–87

CHAPTER VI.

WINTER STATIONS OF ITALY.

BORDIGHERA—Palm grove—Temperature. SAN REMO—its situation and shelter—Vegetation—Climate—Comparison with Mentone—Future winter stations—Dianoalassio—Eastern Riviera—Nervi and Promontory of Ruta—View of Gulf of Genoa—Santa Margherita—Chiavari—Spezia. PISA—Its unsheltered position and moist climate. ROME—Climate—Fevers—Twofold origin of malaria—Parts of city attacked—Connection with depopulation—Causation of malaria—Sun's influence on porous and absorbent soil—Rapid evaporation—Its relation to malaria—Dr. Topham's theory—Rome unsuited to invalids—Objections to large Italian cities. NAPLES—Its drainage—Capri—Salerno—Amalfi—Sorrento, a summer retreat—Influence of the Mediterranean 88–108

CHAPTER VII.

OTHER MEDITERRANEAN WINTER STATIONS.

PAGES

Corsica—Its lofty ranges and sheltered harbours. Ajaccio, a well protected winter station—Meteorological observations—Intermittent fever prevalent in summer—Dr. Bennet's explanation. Malaga—Shelter afforded by double rampart of mountains—Prevalent winds—Small rainfall and high mean temperature—Bad drainage. Tangiers — Great fertility — Climate tempered by Atlantic—Excellent food. Algeria—Climate greatly influenced by Atlas mountains and Sahara desert—Large rainfall owing to prevalence of moist winds—Marshes and fevers. Algiers—Sahel hills—Plain of the Metidja—Hot and rainy seasons—Unequal distribution of rainfall in the three provinces—Its cause—Climate of Algiers compared with that of Riviera — Diseases for which it is suitable—Hill stations   109-125

APPENDIX.

Deep-sea temperatures—Summer stations—St. Dalmas de Tende, Dr. Battersby's report—Certosa di Pesio, Dr. Daubeny's report—Upper Engadine; its extremes of temperature — Danger to Invalids — Hotels — Other summer resorts—Nuovi Bagni of Bormio: its advantages and beauties—Le Prese—Eligible Alpine sites in Switzerland, with Hotel accommodation—The Engadine as a winter residence for consumptive invalids. . 127-131

APPENDIX II.—1870.

FURTHER OBSERVATIONS ON SUMMER QUARTERS FOR INVALIDS, AND ON ALPINE SANATORIA IN 1870.

	PAGE
Superior salubrity of mountain districts—Unhealthy aspect of inhabitants of low valleys—Alpine resorts arranged according to height:—I. Low valleys—Borders of lakes relaxing — Often malarious — Examples. II. Moderate heights, from 2000 to 4000 feet, well suited as summer quarters for invalids—Glyon—St. Cergues—Champery — Sepey — Comballaz — Ormond Dessus—and others in the Diablerets district—Anderlenk—Zweisimmen — Engelberg—Sonnenberg—Grindelwald—Mürren—Rosenlaui, &c. III. High level resorts, from 4000 feet upwards—invigorating effects—Cautions necessary in selection—Curative power of high altitudes in Pulmonary Consumption and Scrofula—Dr. Archibald Smith's observations in Peru, &c.—In Switzerland, &c. Observations of Drs. Lombard, Brehmer, Kuchenmeister, and Hermann Weber — Required heights varying with latitude—Dr. Weber's cases—Question as to the suitability of the mountain climate in winter—Notes on Alpine summer quarters for invalids, by Dr. C. J. B. Williams—Thusis—Journey by Schyn Pass and over the Julier to the Engadine—Severity of the climate in August—Disagrees with some visitors—Highly invigorating to others—Caution and discrimination necessary — Consumption cured in the Engadine — Favourable circumstances and cases—Choice of situation—St. Moritz considered—Samaden, &c.—Tarasp—Davos—Journey in search of another high sanatorium—Over the Bernina—Le Prese—Up the Valtelline—Bormio, new and old Baths—Situation and climate described — Temperature, dryness, &c., compared with other high places—Establishment of the New Baths—Suitable for summer and winter—Thermal waters and baths—Beautiful neighbourhood — Stelvio — Santa Caterina— Routes	135

	PAGES
to Bormio—Other mountain sanatoria—Rigi Kaltbad and Scheideck—Leukerbad—Courmayeur—Gressoney St. John — Monte Generoso — Lauenen —Evolena— Other high resorts for travellers, not well suited for invalids: Andermatt, Furca, Splugen, &c. — Other mountain inns not sufficiently accessible: Æggischhorn, Riffel, &c.	135–161
Concluding Remarks	161–162

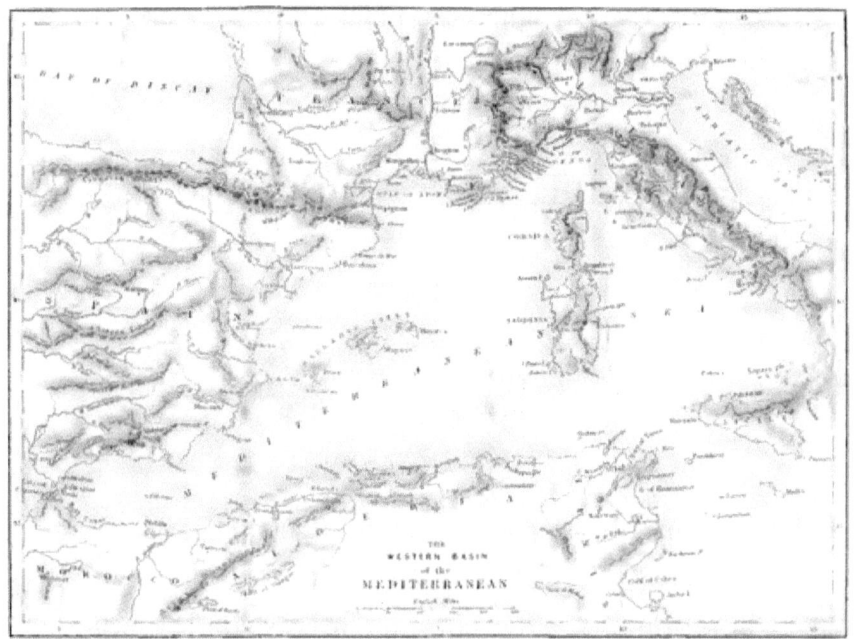

THE CLIMATE

OF THE

SOUTH OF FRANCE.

CHAPTER I.

GENERAL ADVANTAGES.

The sunny Mediterranean shore of France has long been recognised as a fitting winter residence for invalids afflicted with chronic diseases of the lungs; and of late years the portion situated east of Toulon, including the newly annexed department of 'Alpes Maritimes,' has been preferred on account of the shelter afforded to it from northerly winds by the mountain ranges of the Maures, the Estrelles, and the Maritime Alps. This region, which is limited eastwards by the Italian frontier, and includes the towns of Hyères, Cannes, Nice, and Mentone, is remarkable for the luxuriance and semi-tropical character of its vegetation, and enjoys a climate unequalled, within the same latitudes, in Europe, and unsurpassed in the more southern latitudes of Spain and Italy. Its flowers furnish London and Paris with the greater portion of the perfumes used in those cities; while its olives produce an oil of the purest quality, and on that account preferred for pharmaceutical purposes.

The able works of Sir James Clark, Dr. Edwin Lee, Dr. Henry Bennet, and many French writers, have rendered this region comparatively familiar to English physicians; and the completion of the railway from Paris to Nice has made it so easy of access, that it has become the most popular wintering place in Europe.

The further extension of the line to Genoa, and its junction by means of the Cenis and Brenner passes with the countries north of the Alps, will probably considerably increase the number of visitors, and tend to develope new wintering places within this sheltered area, and also along the adjoining coast of Italy. This is much to be desired; for though the French portion of the coast is tolerably well known and accessible to invalids, the Italian is more extensive, and embraces, as will be hereafter shown, many localities, well fitted for winter retreats; but these, owing to their greater distance from rail communication, have not as yet reached the same stage of development as their French neighbours. When an unbroken line of communication exists between Genoa and Nice on the one hand, and Genoa and Paris on the other, the Italian sheltered places will be as easy of access as the French; and the invalid will have a larger number of winter refuges to choose from—a choice of much importance, as it will render him more independent of exacting hotel-keepers and villa owners, and at the same time bring these gentry under the wholesome influence of increased competition.

Before examining in detail the elements of the French Mediterranean climate, some of the general advantages of the invalid's life in this region must be noticed. The chief of these is the amount of sunshine which he enjoys for weeks, and even months together, when the sun often rises in a cloudless sky, shines for several hours with a brightness and warmth surpassing that of the British summer, and then sinks without a cloud behind the secondary ranges of the Maritime Alps, displaying in his setting the beautiful and varied succession of tints which characterise that glorious phenomenon of the refraction of light, a southern sunset; while he imparts to the rugged mountains a softness of outline and a brilliancy of colouring which defy description alike by the painter's art and the writer's pen. Owing to this genial influence, not accompanied, as it is in even the most protected of English wintering places, by any sensation of chill or damp, and the chemical effect of which is seen in the tanning of the skin,—owing to the freedom of the climate from rapid and constantly recurring changes of frost, rain, mist, and mild weather, the invalid spends the greater part of the day in the open air, and scarcely knows what confinement within doors means. The exciting causes of his complaint being removed, and the long spell of propitious weather enabling the full influence of the genial atmosphere to act on his frame, his bodily vigour gradually returns, and he finds himself able to enjoy a fair amount of exercise, whether walking, riding, or driving, in a region in which earth, sea, and sky present to his observation phenomena so varied in form, so brilliant in colour, and so wondrous in beauty, that an inexhaustible feast unfolds itself to his astonished gaze, in the enjoyment of which his attention is withdrawn from the contemplation and ofttimes the exaggeration of his own symptoms, and directed to higher and nobler objects.

In the early stages of phthisis, and especially when

the patient is a young or active-minded man, struck down by overwork or sudden exposure, this cheering influence is most beneficial. It is of great importance that while taking the needful care of himself he should not degenerate at an early age into a hopeless valetudinarian; especially as an every day increasing mass of evidence warrants us in believing that under the influence of medicine and climate a large number of these patients gradually recover their health and lead useful lives, and, with due care, lives of no inconsiderable duration. The freedom from restraint, and the liberty of exercise which the equable climate of the south of France allows, give this class of patients an appearance far different from that presented by the same class at English wintering places. While the sun shines, they certainly do not lead an invalid existence, but, on the contrary, act and live like other people, and it is often impossible for a casual observer to detect any difference in them from the healthy. When the sun sets, their invalid life commences, and continues till his morning rays release them from confinement.

In the very advanced stage of phthisis, where the disease is very extensive, and where little hope of recovery can be justly entertained, it is not as a rule advisable to exile patients so far from their home, or to incur the risk of a deathbed in a foreign land. Still on many the cheering influence of nature's beauties in this bright region is not without its beneficial effect. Often, according to a talented author, the pleasure they derive from gazing on fair scenes has a peculiar intensity from their knowledge that its duration is limited, and that the 'pallid king may at any time make that not unexpected knock and summon them away.'

When I was observing the invalid's life in this region, one feature of it struck me as a forcible illustration of the equability of the climate. It was the comparatively rare occasions on which he has to call in medical aid. Of course there are a certain number of patients who require here, as elsewhere, regular medical attendance; but the greater proportion, and especially those in the earlier stages of consumption, after receiving instructions from the local physicians as to lodgings, diet, exercise, and medicine, continue to carry them out regularly for weeks and months at a time, without the occurrence of fresh complications, and therefore without the necessity for fresh advice. When the weather becomes unfavourable, when exercise is curtailed or wrongly indulged in, when excesses of diet are committed, complications arise, which require the active interference of the physician; and fortunately an efficient British staff is at hand, many members of which are themselves living examples of the healing influence of the climate. Patients should never neglect to consult a doctor on their first arrival, as his experience and advice with regard to lodgings, food, &c., are of great value, and may often prevent them from falling into bad hands, or settling in unhealthy localities.

I have drawn attention to the favourable features of the invalid's life in this climate, but the unfavourable should not be overlooked. It must be admitted that the food is, on the whole, inferior to what we are accustomed to in England; the habits of the people are less cleanly; and the appliances for invalids are neither so good nor so numerous as in some of the British wintering places. The dryness and stimulating influence of the climate, in some of the health resorts, occasionally prevent patients from sleeping so long or so soundly as they are accustomed to in England; and the distance of this region from home will always cause the British wintering places to be preferred for advanced cases.

The drainage of the south of France, though better than that of South Italy, is decidedly defective; and, now that the towns are increasing in size, is fast becoming a source of danger to visitors, which, it is to be hoped, the local authorities will take speedy means to remove. At some of the places the cesspool system is used; but as the tanks are not always provided with sufficient water or with traps and other arrangements to prevent the escape of noxious vapours, these are given off, and are often to be detected in the streets, as at Hyères and Mentone.

In other places the houses are drained into the sea, which, being nearly tideless, does not act the same scavenging part as tidal seas like the Atlantic or the British Channel. Added to this the drain-pipes often open (as at Cannes), not directly into the sea, but on to the beach, a few yards from it, and the contents trickle over the intervening space, giving off foul-smelling gases, until they reach the water. At Nice, some of the drains pass over the dry bed of the Paillon torrent, polluting the small amount of water which remains there in winter, and which is used by the inhabitants for washing clothes. These defects of drainage have not hitherto given rise to any serious epidemic; but the few scattered cases of typhoid fever, which have appeared in some of the past winters, can hardly be assigned to any other cause. The feature of the climate, which so strongly recommends it to medical men, viz. its comparative immunity from rain during the winter months, is the cause of occasional droughts, which limit the amount of water at disposal for drainage purposes; and although in some places, as at Cannes and Hyères, the water is brought from a distance, it is not probable that the supply will ever suffice for a complete water drainage system. For these and other reasons 'the dry-earth system' of Mr. Moule highly recommends itself to the inhabitants of this region, especially as it has been found to answer in India and other dry climates. For the efficient working of this plan the chief requirements are—a good supply of fresh dry earth and a demand for the used material for agricultural purposes. There is no difficulty in obtaining dry earth in sufficient quantity, while the want of material for enriching the soil in a country which produces oil and wine is shown by the fact that manure for that purpose has sometimes to be brought from great distances. Earth closets could be easily introduced into villas by proprietors or visitors, and the system would doubtless spread to the neighbouring towns.

We can now proceed to examine the leading aspects of this climate; to notice in what points it differs from our own; what varieties it presents, and to what morbid states of the body each is applicable. For the sake of convenience it will be well to arrange these aspects into two groups—

1. Physical, including Thermometrical, Hygrometrical, and other Meteorological phenomena.

2. Medical—*i.e.* the Effects of the Climate on Health and Disease.

CHAPTER II.

PHYSICAL ASPECTS.

We have to consider under this heading the principal phenomena of the climate, and we will commence with the *Thermometrical phenomena.*

In discussing the mean temperature of the region collectively, it will be advisable to allow it to be represented by that of Nice. This is rather below the figure that would probably represent it; but the difference is small, and will not materially influence our comparisons.

The mean annual temperature of Nice is 59·4° Fahr.; that of London being 50·3° Fahr.; that of Torquay 52·1° Fahr., and of the Cove of Cork, in Ireland, 51·9° Fahr. I have chosen Torquay as a standard of comparison, because it has the highest annual mean temperature in Great Britain; and the Cove of Cork has been selected for the same reason in Ireland. In spring and winter mean temperatures it is only equalled by Penzance, all the other English wintering places having much lower standards. The mean autumnal temperature of Nice is

61·6° Fahr., compared with 51·3° Fahr. of London, 53·1° Fahr. of Torquay, and 52° Fahr. of the Cove of Cork. The winter *mean*, which is of the most importance to invalids, is 47·8° Fahr. at Nice; while that of London is 39·1° Fahr., that of Torquay 44° Fahr., and of the Cove of Cork 44·1° Fahr. The *mean* in spring is 56·2° Fahr., compared with 48·7° Fahr. of London, with 50° Fahr. of Torquay, and 50·1° Fahr. of the Cove of Cork.

It will be seen from these numbers that this region enjoys a great superiority in point of annual, winter, spring, and autumnal mean temperatures over any spot in the United Kingdom. It has a mean annual temperature of 9·1° Fahr. higher than London, and 7° Fahr. higher than Torquay or the Cove of Cork. But the winter *mean* is the one we have chiefly to do with; and in this point the Mediterranean region has the advantage by 8·7° Fahr. over London, by 3·8° Fahr. over Torquay, and by 3·7° Fahr. over the Cove of Cork.

The contrast between the winter climate of this region and that of England was seldom more marked than during the season 1866–67. On examining the Hyères thermometrical observations, which are most carefully registered by Dr. Griffith, the resident physician, I found that the visitations of terribly cold weather, and the extreme fall of temperature, which prevailed in England and Northern Europe, never penetrated to this corner of the south of France. Subjoined is a comparison of the 'maximum' and 'minimum' Fahr. temperatures at Kew and Hyères during the most severe periods of the winter :—

		Minimum.			Maximum.	
1867.		Kew.	Hyères.		Kew.	Hyères.
January 2	...	19·9	50·0	...	28·7	54·0
3	...	5·7	52·0	...	28·6	55·0
4	...	5·0	54·0	...	16·4	57·0
5	...	1·0	52·0	...	30·8	56·0
14	...	10·5	58·0	...	29·8	65·0
15	...	14·3	56·0	...	31·3	64·0
March 17	...	27·3	53·0	...	36·9	67·0
18	...	11·5	55·0	...	32·7	67·0

It will be seen from this table how marvellous the difference between the 'minima' was, amounting on one occasion to 51° Fahr.; and also what a small amount of variation occurred in the minima at Hyères. Throughout the entire winter the temperature, on only two occasions, fell as low as 32° Fahr., and never below that point; and the weather was, as usual, sufficiently warm and mild to admit of patients sitting out in the open air for the greater part of the day. At Nice the military band played twice a week in the public gardens; and at each performance upwards of 1000 chairs were occupied by visitors and invalids listening to the music. Balls were held in January on board the American men-of-war anchored at Villafranca, and were attended by numbers of ladies and gentlemen from Nice, who were conveyed to and from the ships in open boats.

What are the causes of this great superiority in mean temperature? Let us examine the sources of warmth in the British and French mild climates respectively.

In the case of the British the chief source assigned is the Gulf stream,—which, although it raises the temperature of these islands, and, as Dr. Tyndall has well shown, prevents the formation of gigantic glaciers, has the disadvantage of saturating our atmosphere with moisture, and of adding considerably to the amount of rainfall.

In the south of France the chief sources of its superior warmth are:—

1st. The sun pouring down his warm rays, unchecked in their effects by chill blasts, and rarely obscured by clouds or mists, from sunrise to sunset. The patient arriving in the south of France in winter, sees to his surprise the dust lying on the road; the hills and valleys not bare, as in the land he has lately left, but clothed with evergreens of varied tints, and decked with spring flowers. He sees the walls and rocks apparently alive with lizards, and the air teeming with butterflies and dragonflies, as in summer. His own sensations give him the idea that it is June, and not January, and he forthwith acts accordingly, and dispenses with his overcoat and respirator. For seven or eight hours of the day he can enjoy sunshine, and take walks and rides, or sit on the ground, basking in the warm rays like a lizard. Perhaps this last strikes him more than any other sensation, that he can sit or lie on the soil, for hours at a time, without the slightest sensation of chill or damp—a proceeding which he could never attempt with safety during any winter month in England. He is soon, however, reminded that it is winter by the shortness of the day, the sun setting and night closing in very rapidly. After sunset the temperature falls considerably, as in all countries where the sky is clear; because the radiated heat of the cooling earth is not reflected back to it by layers of clouds, as in England. Heavy dew is not uncommon, and is so white that it is often mistaken for hoar-frost. The freezing point is, however, seldom reached, nor indeed a temperature much below 45° Fahr.; but the lowness of the temperature is not the phenomenon which affects the patient so much as the rapid fall from a high point to a moderate one. This evil effect is easily guarded against. The patient repairs to his room, which, with the windows open, has been basking like himself in the sun all day; and half an hour before sunset the windows are closed; and the heat absorbed during the day is sufficient to keep it at a temperature of 55° Fahr. and upwards, till the next morning. This, I must say, would only apply to rooms facing south or south-west, and not on the ground floor; but the rooms selected for patients are generally first or second floor, and with a southern aspect.* Fires are wanted

* A lady who passed the winter of 1866-7 at Hyères kindly favoured me with thermometrical observations made twice a day for three weeks in an *unoccupied south room*, which had the windows open in the daytime, and was without fire.

1867.		9 A.M.		9 P.M.	1867.		9 A.M.		9 P.M.
February	20	62	...	61	March	4	57	...	56
	21	61	...	60		5	57	...	57
	22	61	...	—		6	56	...	56
	23	62	...	61		7	56	...	57
	24	62	...	62		8	57	...	58
	25	61	...	62		9	58	...	58
	26	61	...	62		10	60	...	64
	27	64	...	63		11	62	...	62
	28	64	...	63		12	62	...	64
March	2	58	...	57		13	62	...	64
	3	57	...	56					

in ground floors, but seldom in upper floors, except during rainy weather.

2nd. The Mediterranean Sea. This body of water presents three important differences from the Atlantic Ocean.

(1.) It has hardly any tide. It is popularly reported to have none at all; but a difference of as much as two feet can be discerned in its most deeply indented bays.

(2.) It contains a larger amount of saline matter than the Atlantic under the same latitude. Messrs. Bouillon Lagrange and Vogel analysed the waters of these seas near Bayonne and Marseilles with the following results:

	Saline residue in 100 parts.
Atlantic, off Bayonne	3·80
Mediterranean, off Marseilles	4·10

The results of sea-water analyses admit of great variations dependent on the depth from which the water is taken; on the latitude of the locality; and its nearness to, or distance from, the mouths of large rivers. As far as can be ascertained, these points were taken into consideration by the above-named eminent chemists: for Marseilles lies in nearly the same latitude as Bayonne (Bayonne is slightly more northern); the water was taken from the same depth in both cases; but the saltness of the Mediterranean is probably reduced at Marseilles by the large volume of fresh water which the Rhone brings down into the sea near that city. The above analyses assign an excess of saline matter to the Mediterranean* in the ratio of 41 to 38 in 1000 parts, or 3 per 1000, and were it not for the last consideration the excess would probably be greater. This extra saltness arises from loss of water by the rapid evaporation from the surface of the Mediter-

---

* A comparison between the analyses of the Mediterranean and the British Channel is also interesting; as it displays this excess in a still more marked degree, and points out to what constituents it is chiefly due.

Amount of saline matter in 1000 grammes of each sea.

	British Channel. (Schweitzer.)	Mediterranean. (Laurent.)
Chloride of sodium	27·06	27·22
Chloride of potassium	·76	·01
Chloride of magnesium	3·67	6·14
Sulphate of magnesia	2·29	7·02
Sulphate of lime	1·41	·15
Bromide of magnesium	·03	·00
Carbonate of lime	·03	·20
	35·25	40·74

The excess amounts to $5\frac{1}{2}$ parts in 1000, and is chiefly caused by the Mediterranean containing nearly twice as much chloride of magnesium, and more than three times as much sulphate of magnesia.

ranean, due to its warm climate; a loss which its tributary rivers and streams, pouring in a smaller volume of fresh water than those of the Black Sea or Baltic, fail to replace. From the Black Sea itself, the waters of which are less saline than those of the Mediterranean, a current sets through the Dardanelles; but as it is weak and superficial, it does not materially dilute the saltness of the latter sea; and during the strong westerly gales which occur in autumn and winter, and whenever the Black Sea rivers are very low, a counter-current towards the more easterly sea prevails. But through the Straits of Gibraltar a strong central current flows constantly from the Atlantic at the rate, according to the late Admiral Smyth,[*] of from two to three miles an hour, while two lateral currents of about the same velocity as the central stream—one on the European, and the other on the African side, ebb and flow, setting alternately with the Mediterranean and with the Atlantic. These lateral streams by no means counterbalance the effects of the central one; and the excess of water which the inland sea thus receives disappears by evaporation, while the saline matter remains behind, rendering the Mediterranean salter than the Atlantic.

[*] 'The Mediterranean.'

(3.) At the places under consideration its temperature is many degrees higher than that of the Atlantic under the same latitude. This is so remarkable a phenomenon, and exercises such an important influence over the climate of the region we are considering, that it may be well to advert to its cause. The Mediterranean is naturally divided into three basins—the western, bounded eastwards by the Adventure and Medina banks, which lie between Sicily and Africa; the eastern, and the basin of the Greek Archipelago, the last two being separated by a ridge 200 fathoms deep. According to the careful soundings of Captain Spratt* these basins are very deep, the western being 9600 feet, and the eastern as much as 13,800 feet. But the most extraordinary phenomenon is the temperature, which at the depth of 100 fathoms in the eastern and western basins amounts to $59\frac{1}{2}°$ Fahr., whereas, according to the researches of Ross, Belcher, Denham, and Pullen, that of the Atlantic,† at the same depth and in the same latitude, is only $39\frac{1}{2}°$ Fahr., shewing a difference of 20 degrees between the two seas.

The temperature of the Greek Archipelago basin is $55\frac{1}{2}°$ Fahr., slightly lower than that of the other two basins; probably owing to the admixture of the

* 'Travels and Researches in Crete.'   † See Appendix.

current from the Black Sea. This great difference from the Atlantic is accounted for by the existence of a submarine barrier of rock extending, according to Admiral Smyth,* from Cape Spartel to Cape Trafalgar, a distance of only 22 miles, and having a width of from 5 to 7 miles. The deepest soundings along this ridge, which the French surveyors have found to be near the Tangiers side, do not exceed 167 fathoms; and it forms a parting wall by which the colder and heavier waters of the Atlantic are prevented from invading the Mediterranean.† Captain Spratt also notices, as a peculiarity of this sea, that it is less influenced than the Atlantic by the alternations of the seasons. The changes of temperature consequent on the seasons do not embrace more than a fall of 10 degrees in winter, and a rise of from 10 to 20 in summer; and these changes only extend to the depth of 100 fathoms, instead of to 500 fathoms, as in the Atlantic of the same latitude, and of 1300 as in the tropics. The Rev. R. D. Graves' observations on the temperature of the Mediterranean off San Remo show that, in the months of November, December, and January, at a depth of four feet, it was never below 60° Fahr. in the daytime.

The first and second points of difference probably

* Op. cit.   † Lyell's 'Principles of Geology,' Vol. I., p. 563.

exercise some slight influence over the temperature of this region; but it is to the third, viz. to the superior warmth of the Mediterranean, that its shores owe their comparative freedom from the depressions of temperature so common in the neighbouring inland places. This sea is an important auxiliary to the temperature of its shores; whilst, by its equalizing influence, it obviates the bad effects of clear skies, and often prevents its shores from sinking below the freezing point through nocturnal radiation. Its superior warmth is explained by Dr. Bennet by the absence of polar currents, whose entrance is prevented by the rocky barrier which exists at the Straits of Gibraltar; but a probable cause of the semi-tropical climate of the south of France is to be found in the tendency of an inland sea to equalize the temperature of all its coasts—a tendency evidently favourable to the northern coast of the Mediterranean.

The beneficial influence of this sea on vegetation cannot fail to attract the attention of travellers. The hills which slope to its warm levels are richly clad with trees and shrubs, their luxuriance increasing the nearer they approach the sea, until they attain their largest size on the shore itself, where, as at Bordighera, even palms are to be seen growing.

This can hardly be accounted for by a southerly aspect, as it occurs where the hills face in other directions, and in places where the shore is comparatively flat, as may be well seen along the Maremma coast between Leghorn and Civita Vecchia. Here the myrtle, the heath, and the pine flourish abundantly, but it is in the thickets which fringe the shore that the largest and finest specimens are to be found. The belts of stone pines on the shore at Cannes, Hyères, and in the Villa Chigi, near Ostia, and the famous Pineta of Ravenna, are like instances of luxuriance induced by the same warming and beautifying influence.

Hygrometrical phenomena.—These are very important in themselves, independently of their influence on the temperature. First, let us consider the rainfall.

The general law of the rainfall in a country is, that, other influences being equal, the *annual fall* of rain is *greater* the *nearer* the country lies to the *equator*, and the *number* of *rainy days* is *greater* the nearer the country lies to the *poles*.[*] Thus, at St. Petersburgh, the mean annual rainfall is 17 in., and it is distributed over 170 days; while, at the

[*] Scoresby Jackson's 'Medical Climatology,' p. 19.

equator, where the mean amount is 95 in., the whole is precipitated in 80 days.

It would be expected that, in accordance with this law, the rainfall in the south of France would be greater than in Great Britain, but such is not the case. The average annual amount at Nice is 25 in., nearly the average at Greenwich, smaller than that of Torquay, which is 28 in., and that of Penzance, which is 44 in.

The law holds good with regard to the number of rainy days, which differs vastly in the two climates. The 25 inches fall in 155 days at Greenwich; whereas the same amount falls in 70* days at Nice. The rain comes down generally in very heavy showers, sometimes as much as $4\frac{1}{2}$ in. in 10 hours; whereas the heaviest rainfall during 24 hours on record at Greenwich is $2\frac{1}{2}$ in. It is manifest that this diminution in the number of rainy days, and therefore increased facility for taking exercise, must be of considerable benefit to patients, as also the decreased amount of moisture in the atmosphere.† This is ascertained by the hygrometer, of which there are several forms,

* Valcourt. Roubaudi states the number to be 60.

† It has been found that the influence of the sun's direct rays in raising the temperature of objects is augmented by the presence of a *limited* amount of aqueous vapour, which obstructs the escape of their heat. Thus, Schlagintweit (Proceedings of Royal Society,

but the one generally used is the wet bulb hygrometer. From observation made on this instrument at Mentone by Dr. Henry Bennet during the winter months of 1864–65, it appears that the average difference of temperature between the wet and dry bulbs was 6·7° Fahr. At Hyères, during the same period, my friend Dr. Griffith found it to amount to nearly 5·0° Fahr. At the Military Hospital at Nice the carefully-recorded figures of Dr. Cabrol (kindly forwarded to me by Dr. Beaugrand), for the same period, show a mean average of 4·92° Fahr. The Military Hospital is situated at some distance from the sea, and near Carabaçel; and it is probable that the average difference in the town of Nice would be greater still.

Contrast these observations with similar ones made at Kew during the same winter months, where the average difference was only 1·46° Fahr.—i. e. less than one fourth of the difference at Mentone, and less than one third of the difference at Hyères and Nice. I may here remark that the Mentone average

March 1865) found that the difference between the temperatures of thermometers exposed in sun and shade in India was much greater, cæteris paribus, in localities on its moist seaboard, than in those of its arid interior; and the scalding influence of the sun's rays between heavy showers, a commonly observed phenomenon, may be attributed to the same cause.

difference for the *winter* of 1864-65 was actually greater than the average difference at Greenwich and Kew during the *summer*.

From this wonderful dryness of the atmosphere there results an almost total immunity from fogs.

We now come to *other meteorological phenomena*. Under this heading I have arranged certain phenomena, which cannot be discussed as purely thermometrical or purely hygrometrical, though they have an important bearing on both.

Firstly, the prevalence of certain winds. The region which we are considering is protected from the north wind, and, with the exception of Nice, from the north-east or *bise* wind; but it is influenced to a certain extent by the *mistral* or north-west wind, to which we will now turn our attention.

The mistral is a wind of prodigious force: often sufficiently strong to blow a man off his horse. It occasionally overthrows the largest trees, and spreads destruction among the corn and vine crops. The trees generally have their branches twisted in the direction of its current; and in the places most exposed to its blasts, screens of wood and stone are erected to protect the vines. It sweeps over the passes of the Maritime Alps, rushing with great fury down the unprotected ravines which lead to the sea, upsetting in its course carriages, carts, and even heavy diligences, and lashing the dark blue waters of the Mediterranean into continuous foam for some distance from the shore.* It is for the most part a dry wind, parching up the country, and withering the leaves of plants by its desiccating influence. The barrenness of the low ranges of mountains near Marseilles may be attributed to this cause. Its dryness is shown by its effect on the hygrometer, which sometimes, during its prevalence, indicates a difference of as much as 10° Fahr. between the bulbs.

* In January 1868 I travelled with three ladies, in a four-horse carriage, over the Bracco Pass, between Spezia and Genoa, when the mistral was blowing. On arriving at the Col the wind broke a window and tore the leather off the top of the carriage. The two leaders were blown round to the very edge of the road, which is here flanked by a precipitous wall devoid of parapet. The horses stood cowering and scarcely able to retain their footing against the force of the blast; but the coachman and I, after a hard struggle, in which it was a question who should win, the mistral or we, succeeded in turning the horses slightly away from the precipice; the ladies were got out of the carriage, and, with the help of some men who came up, the leaders were unharnessed. The carriage then, assisted by drags and sundry men holding on to the wheels, proceeded to the next town, while our party followed on foot. This was no easy matter, for the wind swept over the Col with such violence, that we were blown down several times before we succeeded in reaching the more sheltered part of the road. On our way we saw a cart, laden with hay, laid flat on its side; and we learnt that in the previous year the diligence had been upset, and that accidents of this description occasionally happened during the months of January and February.

Various theories have been started as to its origin. M. Martins considers that it is the result of the denudation of the Rhone basin by the destruction of its forests; and he alleges that in the time of Julius Cæsar, who described this country as covered with forests, this wind did not exist. 'When,' says M. Martins, 'the denuded crests of the mountains and the hilly plains become heated by the sun's rays, the air in contact with them, being likewise heated, dilates and rises into the higher regions like that from a chimney where a strong fire is burning. The colder and heavier air which surrounds the snowy summits of the Alps precipitates itself in an aërial torrent to fill up the partial void which has been occasioned by the ascent of the lower stratum of air. Were the country covered anew with forests the causes productive of the mistral would be in a great measure removed.' It will be observed that this theory represents the mistral as a cold current coming from the Alps; but unfortunately the Alps lie *north-east* of the district where the mistral is most felt, and an aërial current coming from them would therefore give rise to a north-*easterly* wind, and *not* a north-*westerly*.

Dr. Bennet considers that it is sometimes a local wind, originating in the south of France and causing a clear sky; at other times a grand north-west European wind, coming from the North Seas and North-West Atlantic; and he states that then it occasionally, though rarely, brings with it black clouds. With reference to this view I may observe:—

Firstly. That the gigantic force of this wind cannot be satisfactorily accounted for by any theory assigning it a merely local origin. The difference in temperature between the region to the north and west of the parts ravaged by this wind is quite insufficient to account for its powerful blasts.

Secondly. That any wind passing from the North Seas to the region under consideration would be a north-east wind, since the effect of the earth's rotation upon a wind blowing from the North Pole to the Equator would be to convert it into a north-east wind.

I therefore feel bound to reject the explanations hitherto given of this wind's origin as unsatisfactory.

In France, the mistral is only known in the parts lying near the Mediterranean coast, and to the east of the Cevennes range of mountains; for north-west winds experienced in other parts of that country are of an entirely different nature, being more or less moist, and not possessed of great force. Therefore, as it has been shown that the mistral cannot have a mere local origin, it exists probably as an upper current and descends on the region under consideration. And this is the more probable, because the formation of powerful upper aërial currents is attended with less difficulty than that of surface winds, in consequence of their not being subjected to friction with the earth's surface. It will be seen that the foregoing conclusion affords a satisfactory explanation of the *dryness* of this wind. For even if the upper atmospheric current were saturated with moisture, yet the amount of vapour held in suspension must be small on account of the lowness of the temperature; and when that temperature is raised by the compression which follows its descent into the lower atmospheric regions, the aërial current must be far from a state of saturation. In fact, as the mistral must come from the direction of the Atlantic, I do not see how its dryness can be satisfactorily accounted for in any other manner. But why does such a current descend upon this particular region? Doubtless because the air over the warm Mediterranean basin becoming heated and rarefied, a partial vacuum is produced, into which the cold upper current descends. I think, therefore, I am fully justified in assuming that the mistral must be caused by an upper aërial current descending on this region. It remains for me to explain how such an upper current may be formed.

A glance at the map shows that the south coast of France is in the same latitude with the south of Canada, where the winters are extremely severe, mercury having been known to freeze; and on casting our eyes farther north we come to the cold regions of British North America and the great Arctic Archipelago. There will be no difficulty in accounting for the low temperature of any wind blowing from this quarter. It will also be seen, on inspecting any map of America showing the lines of equal mean January temperature, that the coldness of the British North American climate increases rapidly as we proceed westwards into its interior, such increase of coldness being fully as great as what would be encountered in proceeding from south to north. The intense cold of the interior parts of this region tends to produce an atmospheric current outwards in a direction at right angles to these isothermal lines. This will be found to be a direction from west-north-west. Such a current, which would probably extend to a considerable height in the atmosphere, would be raised in altitude by encountering the high lands of Labrador, lately surveyed by Professor Hind; after which, it, or at any rate the upper portion of the current, might continue its course at a height above the Atlantic, crossing the North Polar current and the Gulf stream as an upper atmospheric current.

But, it may be asked, why should not this current descend to supply the partial vacuum caused by the heating and rarefaction of the air in contact with the Gulf stream? Such a descent is rendered improbable by the nature of the stratum of air overlying this warm stream. It must be loaded with aqueous vapour. Now, the experiments of Dr. Tyndall show that the heat-absorbing powers of aqueous vapour are enormous; that although in pure atmospheric air its amount in atoms is only 1 to 200 of oxygen and nitrogen, the single atom of aqueous vapour absorbs more heat than the 200 atoms of oxygen and nitrogen collectively, and, compared with the action of a single atom of oxygen or nitrogen, its heat-absorbing power is 16,000 times as great. The stratum of air in contact with the Gulf stream is, as has been stated, loaded with aqueous vapour, and therefore is possessed of absorbing properties sufficient to form a screen nearly impervious to the heat radiated from the warm body of water. Owing to this, and also to the feeble radiating power of water, the decrease in temperature in ascending to the upper strata of the atmosphere overlying this stream must be much more rapid than in ascending into atmospheric strata overlying land. Such being the case, it is clear that the rarefaction produced by the warmth of the Gulf stream would not extend to any considerable height in the atmosphere. Moreover, the partial vacuum before mentioned is supplied by the upper equatorial current, which here descends and forms the south-west wind prevalent in the temperate regions of the Atlantic. The west-north-west current would, therefore, continue its course as an upper atmospheric current little affected by the warmth of the Gulf stream, and could not hold any great amount of moisture by reason of its low temperature. This current would, if not acted upon by disturbing influences, cross the Atlantic as an upper west-north-west current. But as the effect of the earth's rotation upon a north wind in the northern hemisphere is to give it an easterly tendency, and transform it into a north-east wind, so this easterly tendency, in the case of a west-north-west current, would convert it into a north-west current. This north-west current would, as before pointed out, descend upon the warm Mediterranean basin as a dry north-west-wind. Hence the mistral.

It appears from hygrometrical observations, that

the mistral, although dry, is not so dry as the bise or north-east wind prevalent at Nice, and in some other parts of the south of France—a circumstance which is evidently in complete harmony with this theory, as the aërial current in passing above the Atlantic must receive some vapour; and although not sufficient in amount to prevent its being a dry wind for the reasons before indicated, yet we should expect to find it moister than the north-east wind, which must have crossed the Alps before its arrival at Nice and other southern places.

I am aware that this theory would require a considerable amount of evidence to establish its truth; but it appeared to me reasonable, as it suggests a cause worthy of so powerful an effect as the mistral, by assigning to it a cosmical origin. This is analogous to the explanation given of other powerful currents, as the monsoons and the south-west equatorial current, &c. However, as yet I can offer it only as an hypothesis, the truth of which must be tested by subsequent observation.

The southerly winds, which prevail in the south of France, are for the most part moist, and generally accompanied by clouds and rain. Beyond their bringing saline breezes and occasionally boisterous weather, they are not injurious. An exception, however, is the south-east, a sirocco wind, which is in general warm, and in the more sheltered watering-places occasionally causes a decidedly oppressive state of atmosphere. At Hyères, it is annoying, because it blows into the town the dust of the plain intervening between Hyères and the sea. It is, however, a moist wind, and rain soon follows to lay the dust. Dr. Bennet describes this wind at Mentone as having 'all but lost the languor-giving properties which distinguish it at Naples.' He states that it sometimes becomes exceedingly cold in winter from the presence of a large amount of snow on the mountains of Corsica; and that it may even cause a fall of snow at Mentone, which occurred in the very severe winter of 1863-64. My own experience of this wind at Nice and Mentone is, that it is moister than in South Italy, but at times very warm and enervating. The east wind, which is scarcely felt except at Nice and Cannes, is less dry than in England; and the west wind retains its usual character of greater humidity, but is not so loaded with moisture as in England.

Secondly. The climate is considerably influenced by the neighbourhood of mountain ranges. This is an important feature of this climate, as it is through this region being situated to the south of mountains that it owes its general protection from northerly winds. There is an important point to be considered when we attempt to judge of the shelter which a mountain to the north of a town affords— namely, whether the town is situated immediately *under* the mountain or *at some distance* from it. This constitutes the chief cause of difference between the climates of Nice and Mentone. Nice is situated upwards of three miles from the base of its projecting chain, and therefore does not enjoy the same immunity from northerly winds that Mentone does, which lies immediately under an amphitheatre of mountains.

The health resort should be situated immediately under, or on the lower slopes of, the protecting range, which should be wooded to the summit, and not sufficiently lofty to allow snow to rest for a long period in winter; or the streams which descend its sides will be reduced in temperature, and bring down currents of cold air with them. Hyères is greatly favoured in these respects, as it is sheltered by wooded hills of no great height, and lies at a considerable distance from snow ranges and mountain torrents. Cannes is much nearer to the snow-capped mountains; but the adjoining hills are well wooded, and few streams reach its valley. Nice is unfortunately influenced by the torrent Paillon, which is cooled by the glaciers of the Col de Tende range: and Mentone, though enjoying excellent shelter, is intersected by torrents which, descending from the Maritime Alps, traverse its protected area.

CHAPTER III.

HEALTH RESORTS—HYÈRES—COSTABELLE—CANNES NICE—CIMIEZ—CARABAÇEL—MENTONE.

The principal meteorological phenomena presented by the climate of the whole region having been considered in the last chapter, we can now direct our attention to the modifications of it found in the individual health resorts; and these will be described briefly, but faithfully, in the order in which they are approached by rail from Paris. The first arrived at, after one hour and three quarters' journey from Marseilles, or nineteen hours from Paris, is *Hyères*, a town of 10,000 inhabitants, situated in latitude 43·7° N., and the nearest and the most southerly of all the French winter resorts. The town is built on the most southern slope of one of a range of hills called the Maurettes, and faces south and south-east. It is distant about three miles from the Rade d'Hyères, an inlet of the Mediterranean, the intervening space consisting of a very fertile plain. The valley of

Hyères runs in a north-westerly direction back from the sea, between the well-wooded ranges of the Paradis and the Maurettes, and opens on to the town and harbour of Toulon, another inlet of the Mediterranean, being to a certain extent closed in by the picturesque mountains over that seaport. On approaching Hyères from Toulon, a sensible difference, as Dr. Edwin Lee justly remarks, is experienced in the temperature, owing to the southern portion of the valley being more sheltered; and its increased fertility is shown in the extreme luxuriance of the vegetation. The ranges of the Paradis and Maurettes, instead of being bare, like the majority of the mountains of Provence, are clothed to their summits with pine and cork trees. In the valley abundant crops of corn, wine, and oil are produced apparently on the same spot of ground, the wheat and the vine crops attaining to maturity even under the shade of the olive trees. Oranges and lemons are neither so fine nor so abundant as at Nice and Mentone; but the aloe and cactus, common throughout this region, grow in great luxuriance and profusion at Hyères. The striking feature of the vegetation, and one which imparts a tropical aspect to the landscape, is the presence of the palm-tree in numbers and rare beauty, attaining a greater height and a finer growth here than in any other part of France. In M. Denis's garden are to be seen no less than fifteen different varieties of this tree, many of which are natives of the Cape of Good Hope, flourishing in the open air. The date-palm, *Phœnix dactylifera,* is the most common; but other varieties, including the fan-palm, *Chamærops Fortunei,* are not wanting. M. Denis exhibited some fine date-palms in the Jardin Réservé of the Paris Exhibition (1867); and one of them was of remarkably vigorous growth, and had growing around it young ones self-sown. The wild flowers are abundant, and of varied and brilliant colours. According to Mr. Timins's observations, they bloom somewhat earlier here than at Cannes.*

* The Rev. D. C. Timins, M.A., Oriel Coll., Oxon., has kindly favoured me with the time of appearance of the following plants in bloom at Hyères and at Cannes respectively. These observations were made by Mr. Timins, at Cannes, in 1865, and at Hyères, in 1866.

	Hyères.	Cannes.	Difference.
Anemone hortensis	Jan. 17	Jan. 26	9 days
coronaria	Feb. 6	Feb. 17	11 ,,
pavonina	Feb. 16	March 17	29 ,,
Ranunculus ficaria	Jan. 22	Feb. 13	22 ,,
Narcissus tazetta	Jan. 30	Feb. 14	15 ,,
Cerintha aspera	Jan. 30	Feb. 12	13 ,,
Linum flavum	Feb. 6	Feb. 13	7 ,,
Tulipa oculus solis	Feb. 15	March 1	14 ,,
Muscari cornosum	Feb. 14	Feb. 28	14 ,,
Erica arborea	Feb. 2	March 8	34 ,,
Arisarium vulgare	Jan. 27	Feb. 14	18 ,,

The Iles d'Hyères, the ancient Stœchades, lie off the coast, and, with the Presqu'île de Gien, help to form the Rade d'Hyères. These islands are hilly and precipitous, and to a certain extent screen the

	Hyères.	Cannes.	Difference.
Arum italicum	March 10	April 5	26 days
Iris germanica	Feb. 15	April 11	24 ,,
pseudacorus	Feb. 15	April 11	24 ,,
Helianthemum album	March 30	April 5	6 ,,
polyfolium	Feb. 28	March 15	15 ,,
Gladiolus germanicus	April 12	April 15	3 ,,
Borago officinalis	Jan. 31	Feb. 14	14 ,,
Vinca minor	Jan. 17	Feb. 7	21 ,,

In order to judge better of these comparisons, I have compared the temperature of the two seasons at Cannes and other places; and although I find that the season of 1865 was, on the whole, colder than that of 1866, the difference of mean temperature in the month of February amounting to as much as 4° Fahr., yet the difference of the seasons is not, I think, sufficient to account for the striking contrast presented by Mr. Timins's table; more especially as the mean temperature of April 1865 was actually 6° Fahr. higher than that of 1866. It will be seen that, out of twenty plants, two flowered at Hyères a month and upwards earlier than at Cannes; six, three weeks and upwards; nine, one week and upwards.

Mr. Timins has also furnished me with a list of the principal species of Lepidoptera, and their time of appearance at the two places taken during the same seasons as the plants. It is here subjoined, as it may probably interest some of my readers.

Species.	Appeared at Hyères.	At Cannes.
Papilio Machaon	March 25	April 7
Podalirius	Feb. 26	April 8
Thais Cassandra	March 27	April 5
Thais Medecicasta	April 12	April 10

town of Hyères from the sometimes rather boisterous sea winds. Their appearance, as seen on a fine day from the Place des Palmiers, is strikingly beautiful; their fine outlines rising hazily from the blue

Species.	Appeared at Hyères.	At Cannes.
Rhodocera Cleopatra	Feb. 13	April 7
2nd brood	Oct. 29	Oct. 10
Rhamni	March 31	April 7
Colias Edusa	Jan. 29	Jan. 27
myrmidone	Feb. 14	April 10
Colias Hyale	Feb. 19	March 12
2nd brood	Oct. 30	None seen
Aporia Crataegi	April 25	None seen
Pieris Belia	Feb. 3	Feb. 7
Ausonia	March 6	March 28
Daplidice	Feb. 5	Feb. 4
Bellidice	Jan. 30	Jan. 27
Cardamines	April 4	April 12
Euphæno	April 8	April 26
Leptosia candida	March 27	April 17
Lathyri	Feb. 28	April
Erysimi	April 15	None seen
Argo Psyche	April 25	None seen
Satyrus Xiphia	Feb. 17	March
Egeria	March 28	April 6
Megæra	Feb. 2	March 1
adrasta	None seen	April 26
Vanessa cardui ⎫	Feb. 8	March
polychloros ⎪ Hibernated	Feb. 8	March 1
Io ⎬ specimens.	Feb.	April
Antiopa ⎪	March 17	April 10
Atalanta ⎭	Feb. 1	Jan. 23
Grapta C. album	April 12	April 10
triangulum	March 11	April 10
Limenitis Camilla	April 12	May 7

Mediterranean, and forming an important feature in the sunny view enjoyed from that spot. On the largest island, Porquerolles, remarkable for the great size and fragrant odours of its shrubs, the French Government have established an Hospital for the invalid and wounded Algerine troops, thus giving a practical proof of their opinion as to the mildness of the climate.

The east end of the town of Hyères is completely sheltered from the mistral by the Castle Hill, a spur of the Maurettes; but the west end is open to its influence. Its force is broken by the mountains over

Species.	Appeared at Hyères.	At Cannes.
Melitæa Deione	April 12	May 6
Cinxia	April 12	April 20
Provencialis	April 12	None seen
Didyma	April 25	None seen
Dia	None seen	April
Argynnis Lathonia	April	Feb. 7
Thecla rubi	Feb. 9	March 18
Lycæna ballus	March 17	None seen
Hippothoö	April 24	April 18
Polyommatus Argiolus	Feb. 28	April 7
Hylas	Feb. 27	April 6
Melanops	April 12	May 6
Bœticus	Oct. 29	Nov. 5
Spring brood	April 24	None seen
Telicanus	Oct. 29	Nov. 5
Amynfas	Oct. 30	None seen
Battus	None seen	April 26
Hesperia alveolus	March	April

Toulon, the Coudon and the Pharon, and its blasts are not quite so dry as at Marseilles, probably on account of the seaboard over which it passes before reaching Hyères. The town is protected from east winds by the chain of Les Maures: but it is affected by the south-east wind, which, as has been before stated, is sometimes irritating from the dust it brings up from the plain below. The Hôtel du Parc and the Hôtel d'Orient are situated at the east end of the town, and are therefore protected from the mistral; the Hôtel des Iles d'Or and the Hôtel des Hespérides lie at the west end, and consequently are exposed to some of its gusts.

The mean winter temperature is 47·3° Fahr.—i.e. lower than that of Mentone, and not differing much from that of Nice. The average annual rainfall is twenty-seven inches—i.e. greater than that at Nice, and probably greater than that at Mentone. The amount of rainfall is curious, and contrasts with that of Toulon, which is eighteen inches, and doubtless this difference, combined with a higher mean temperature, accounts for the great luxuriance of the vegetation in the neighbourhood of Hyères. The number of rainy days, according to M. de Valcourt, is sixty-three—i.e. less than that at Mentone or Nice: but the difference in the amount of moisture is best shown by the hygrometer. During the winter 1864–65, as has been before stated, the mean difference between the bulbs was 6·7° Fahr. at Mentone, while at Hyères it was 5° Fahr.

The climate of Hyères is the least exciting and the least stimulating of all health-resorts of this region. In fact, it sometimes has a sedative effect; for Dr. Griffith, the resident English physician, informed me that many patients coming from Mentone and Cannes in a state of nervous excitement and wakefulness, brought on by the stimulating effect of the air, have slept well and soundly at Hyères. This quality in the climate has its origin partly in its somewhat greater humidity when compared with that of Mentone, as seen above; partly in the luxuriance of the vegetation; and partly in the fact of the town being at some distance from the sea, and to a certain extent screened from saline breezes. Though not so perfectly sheltered from the mistral, nor enjoying so high a mean temperature as Mentone, the openness and breadth of its valley gives Hyères the advantage of a freer circulation of air, and of a larger space of level or gently sloping ground; thus affording greater facilities for exercise in those cases where mountain climbing is unadvisable, and where drives or walks on level ground are preferable.

I must not omit to mention a sunny spot called *Costabelle*, lying one mile from the sea, and about a mile and a half from the town of Hyères. It is completely sheltered from the mistral by the Pic des Oiseaux, a spur of the Paradis range, and, according to M. Denis, enjoys a mean temperature two degrees of Fahrenheit higher than Hyères; but its closer vicinity to the sea renders it more open to the southerly winds. The exotics growing in the villa gardens testify to the mildness of the climate; and as it is a small place, and consists for the most part of detached villas, it possesses advantages in a hygienic point of view over most parts of the town of Hyères, where the houses are crowded, and the drainage arrangements by no means perfect.

Cannes, a town of 8,000 inhabitants, may be reached in about three hours by rail from Hyères, and in one hour from Nice. It is prettily situated in the deepest recess of the gulf of Napoule, opposite the low islands, Les Lerins. From the Pointe de la Croisette, the eastern boundary of the gulf, the wooded hills of Vallauris run back in a north-westerly direction, and protect from easterly winds a fertile plain separating them from the sea. These hills rise to a considerable height, and are continuous with some undulating ranges, situated to the north of the plain, which afford to Cannes its chief shelter from the northerly blasts. The line of protecting ridges, by no means an unbroken one, is completed by the Estrelles themselves, which are separated from Cannes by the wide plain of Laval and the Siagnes stream. Further on they descend in graceful masses into the Mediterranean, forming the western boundary of the Gulf of Napoule, whose calm waters lave their bases, and of whose exquisite scenery their rugged and varied outlines are the great charm. The basin thus formed is a wide one, and portions of it lie too far from the protecting barriers to be completely sheltered by them. These portions cannot enjoy the same immunity from cold winds which others lying immediately under them obtain; and for this reason there is a great difference in the degree of shelter enjoyed by different parts of the Cannes basin. The town of Cannes is divided by the Mont Chevalier into east and west portions, the eastern being situated chiefly on level ground in close proximity to the sea; while the western, rising on the slopes of some wooded heights, is further removed from the shore, and looks towards the Estrelles mountains.

Cannes is well sheltered from the north and northeast winds, and, to a certain extent, from the north-west or mistral, but owing to the depressions in the Estrelles range and the distance of the town from them, this wind at times prevails with considerable power. The hills of Vallauris give some protection from the east wind, which is of an irritating character, though less so than at Nice.

Cannes is exposed to the blasts of all the southerly winds, as the islands, Les Lerins, are too flat and lie too near the sea level to afford efficient protection. The winter mean temperature is 48° Fahr.—i.e. lower than at Mentone, and higher than at Hyères. The average number of rainy days in the year is fifty-two, the smallest number in this region. The annual rainfall is twenty-five inches, the same as at Nice. (I have been favoured with some hygrometrical observations, diligently and carefully recorded by Mrs. R. Cocks, who has passed several winters there. As far as they admit of comparison with those of Mentone, they show Cannes to have a less dry climate.*) The vegetation is very rich and

* During the months of January and February 1865, the difference between the bulbs was 4·8° Fahr. At Mentone Dr. Bennet's observations, taken during the same months, and at about the same hour of the day, showed a difference of 7° Fahr. It is only fair to state that my friend Dr. Frank, after comparing the observations of Dr. de Valcourt at Cannes, and Mr. Freeman at Mentone, made during the same months of 1865, has arrived at a different conclusion from the above.

of varied description, as is seen in the villa gardens; but the chief feature is the prevalence of scent-producing plants. Whole fields are devoted to the cultivation of the jessamine, cassia, and geranium, so that the air is redolent with these perfumes. Another fine feature of the vegetation is the graceful Pinus pinea, or umbrella pine, which is extremely abundant here, and attains to a very great size, probably from the favourable nature of the soil, which is of a sandy character.

The climate is warm, dry, and decidedly stimulating; as might be expected from the situation of the town on the coast, and from the sandy and schistose nature of the soil. Cannes enjoys the advantage over Nice, of being less exposed to sudden changes in temperature, on account of its superior protection from northerly winds. The climate is nearly as stimulating, and contrasts greatly with the sedative qualities of that of Hyères.

Of the numerous and excellent hotels which Cannes possesses, the Grand Hôtel de Cannes, Hôtels Impérial, des Princes, and de la Méditerranée are built on the Boulevard de l'Impératrice, close to the sea, on the eastern side of the town; while, on the same side, but situated at a considerable distance inland, are the Hôtels de Provence and de l'Europe, Victoria, and Beau Séjour.

At the west end is the Hôtel Pavillon, standing a short distance from the shore, while further removed, and occupying elevated positions, are the Hôtels Bellevue and Beausite. The recently completed waterworks now furnish the town with a bountiful supply of good water.

The pretty village of Cannet lies only two miles from Cannes, nestling on the wooded side of Vallauris, and its admirable position at the end of a well-sheltered valley, enjoying a beautiful view of the coast line, seems to render it very suitable for a winter station for invalids. As Cannet promises to supply a want much felt at Cannes, and only partially provided for by the Hôtels de l'Europe and Provence, viz. a well-sheltered inland retreat from the too boisterous sea winds, it is curious that the advantages offered by the spot have not as yet been fully developed.

Dr. Frank, one of the English physicians at Cannes, informs me that villas are now springing up in the neighbourhood; and it is to be hoped that ere long Cannet and the Vallauris district will form a winter station, whose climate, as judged of by its position, would much resemble that of Hyères.

The valley of Vallergues, too, on the road to Grasse, at some distance from the shore, enjoys many of the same advantages as Cannet, being well sheltered and away from the sea.

Nice is a considerable town of some 40,000 inhabitants, situated on the Mediterranean coast between Cannes and Mentone. It faces the south, is built on a plain, formed probably by the deposits of the Paillon torrent; and is surrounded, except on the seaboard, by an amphitheatre of mountains. This plain runs back from the sea to the distance of three miles and upwards; but laterally it occupies a far greater space. The protecting amphitheatre of mountains consists of spurs of the Maritime Alps, forming a succession of ranges, of which the nearer ones are low wooded hills, pleasantly besprinkled with villas, while the more distant are lofty and bare. The higher chain of mountains does not afford by any means complete shelter to the town, partly on account of its distance, and partly on account of various gaps and depressions in the range. The chief gap in the Nicean amphitheatre is to the north-east, and is caused by the wide valley of the Paillon, a torrent, which flows down from the snows and glaciers of the Col de Tende, and discharges itself into the Mediterranean at Nice. The *bise*, or north-east wind, reaches Nice through this valley, and is very cold and cutting. The *mistral* comes through a depression to the north-west, and is more felt here than at any of the four health-resorts. The east wind prevails in March, and is dry and of an irritating nature; and all the southerly breezes blow unchecked by obstacles of any sort. The mean winter temperature is, as has been stated before, 47·8° Fahr. The average rainfall is twenty-five inches, and the average number of rainy days, sixty.

Cimiez and Carabaçel are suburbs of Nice, and from their situation enjoy a much greater amount of protection than the town itself. Cimiez is placed on the slopes and near the summit of one of the low ranges to the north of Nice, while Carabaçel lies immediately at its foot. Carabaçel is upwards of a mile from the sea, is nearly entirely protected from northerly winds by the hill of Cimiez, and but slightly influenced by southerly and easterly breezes. Cimiez is still further from the sea, and enjoys equal protection; and in both these places the vegetation is very luxuriant, and exotics of varied description flourish in the open air.

To the north-east of Nice, at about two miles from the sea, lies the district of St. Barthélemy, the greater part of which is sheltered from the mistral by wooded hills, which rise to the north and west of it. Its distance from the sea, and sheltered position, render its climate similar to that of Carabaçel and Cimiez, and as both these places are overcrowded, it is to be hoped that accommodation for invalids will be forthcoming at St. Barthélemy, and thus the number of our winter refuges will be increased.

The climate of Nice itself is warm, very dry, and rather stimulating; but its chief defect lies in its liability to sudden and rapid changes of temperature, which arise from its imperfect protection from northerly and easterly winds. When snow falls on Mont Chauve, a mountain to the north of Nice, and the wind blows from that quarter, a bitter state of atmosphere is experienced in the town. Although the uncertainty of climate precludes Nice from being suitable for the majority of pulmonary complaints, the absence of moisture, and the combination of bright sunshine and saline breezes, renders it eminently suitable for rheumatic and gouty affections, as also for the atonic forms of dyspepsia and for many scrofulous complaints.

The appearance of intermittent fever in Nice and its neighbourhood, and the increase in the number of cases admitted into the hospital, have naturally drawn attention to the causes of the malady. The Paillon and Var torrents, which formerly desolated a wide tract of country, have been confined within certain limits by aid of embankments, and thus a great deal of land for cultivation has been gained; in which, as is usual in this country, artificial irrigation has been established. The embanking, while it has narrowed the course of the streams, has raised the level of their beds, which in some districts is above the neighbouring land; and it is in these districts that the malarious influence is chiefly felt. Sir Ranald Martin tells me of a similar example in India. Through a plain, previously healthy, a canal was carried at some height above the neighbouring country; and shortly afterwards intermittent fever appeared among the inhabitants dwelling on either side of the canal. Near the Var embankments, a good many cases have occurred, and Dr. Pantaleone, who, during his residence at Rome had great experience of this kind of fever, told me that cases of a very severe type had fallen under his notice, and that in most cases the disease had been contracted by visitors frequenting picnics in this neighbourhood.

A great deal of the land on the right bank of the Paillon lies below the bed of the river; but in the suburbs of Nice the new roads are constructed at a high level, and the ground lying between them is always raised previously to being built on. In course of time, therefore, 'the levelling up' of the country will (unless the beds of the rivers are also raised) abolish this cause of malaria, which in the Paillon district is now only slight.

The climate of Carabaçel and Cimiez presents a decided contrast to that of Nice, and some similarity to that of Hyères, being less exciting, less liable to sudden change, and moister than that of Nice. The hygrometrical observations, taken at the military hospital near Carabaçel, show the amount of moisture to be nearly the same as at Hyères; and several medical friends have informed me that this climate exercises the same sedative influence on patients coming from Cannes and Nice itself, as Dr. Griffith states is exercised by the climate of Hyères. This decided contrast to the climate of Nice, displayed by its suburbs, is to be accounted for by their distance from the sea, by their large amount of vegetation, and by their superior shelter from all winds. Pulmonary invalids may therefore find a safe refuge in these localities, if they are content to remain within their bounds. But this is too often not the case; they complain of being dull and moped in the quiet retreats of Cimiez and Carabaçel; and, tempted by the attractions and promenades of Nice, they encounter the sudden changes and chills which are so prejudicial to such cases, and may frustrate all the objects for which they have left their native land.

The Hôtels Victoria, Méditerranée, Grand Bretagne, Angleterre, and des Anglais, are close to the sea. The Grand Hotel, the Hôtel de France, and Hôtel Chauvin are on the right bank of the Paillon, at a short distance from the shore; while in sheltered positions, at a considerable distance from the sea, are to be found the Hôtel Royal, and the Hôtel du Louvre in the Rue Grimaldi, and the Pension Milliet in the Rue St. Etienne. At Carabaçel, the Hôtels de Nice and Paris, and at Cimiez, the Pension Garin, are tolerably protected from cold winds. The railway at present only extends as far as Monaco, which is reached in about an hour from Nice; and a drive of another hour brings us to Mentone. The exposed situation of Monaco, perched on a rocky peninsula, which is open to nearly all the winds of heaven, unfits it for invalids; but the sheltered and lovely coast road to Mentone, on which there are at present some villas, seems to offer advantageous sites for further building.

Mentone, a town of 5000 inhabitants, situated twenty-two miles east of Nice, and close to the Italian frontier. It is beautifully placed at the foot of some wooded hills, backed by the Maritime Alps, which here rise to the height of 3000 or 4000 feet, and form a semicircle, completely protecting the town from all northerly winds. A fine bay, bounded on the east by the Murtola Point, and on the west by the Cape St. Martin, is divided into two smaller ones, the eastern and western bays, by a peninsula, on which the town of Mentone stands. A number of small valleys run back from the town towards the protecting range; and it is in these, and in the gardens immediately behind Mentone, that the vegetation is so very luxuriant, and that the lemon-tree attains a degree of perfection unequalled elsewhere in France. The fruit falls off the branches if the temperature sinks to 27° Fahr., and the tree itself is killed at 24° Fahr.; so that the appearance of this tree furnishes a tolerably fair indication of the thermometrical changes taking place. Olive and carob trees grow to a great size, and are very productive. The houses occupied by visitors are built for the most part close to the sea, and though a few villas have been erected away from it, the short distance that intervenes between the sea and the protecting ranges will prevent a large number being built, except in the immediate vicinity of the sea. I will not enter into a fuller description of this beautiful and well-protected spot, which has had full justice done it by Dr. Henry Bennet, in his charming work on 'Winter in the South of Europe,' and by Dr. Siordet and other authors. Mentone is completely sheltered from all northerly winds, including the mistral; but it is open to the east wind, and to all the southerly breezes, including the sirocco, which is, perhaps, the only objectionable wind that visits this town, and causes a very close state of atmosphere when it has been blowing for some days. The mean winter temperature is 49·5° Fahr. (Bennet), 48·5° Fahr. (Valcourt), the highest in this region. The average number of rainy days is eighty, and the rainfall is said to be greater than at Nice. The hygrometical observations of Dr. Bennet, as cited above, for the winter 1864-65, show Mentone to be the driest of all the four localities where registered hygrometrical observations are kept.

The climate is warm, very dry, and stimulating. It is also very equable, being much less liable to sudden changes of temperature than Nice or Cannes. There is a want of circulation in the atmosphere, particularly of the eastern bay; and the close proximity of most of the houses to the sea subjects patients too much to the noise and stimulating effects of that element.*

* This defect in Mentone may some day be supplied by Rocca-

Of the two bays, the eastern is the warmer and more sheltered, except from the east wind, the houses being built immediately under the protecting mountain range; but the western bay, though less sheltered from all winds except the east, offers a better chance of inland accommodation, and is sometimes preferred on this account. Dr. Frank informed me that the plateau of Pian, in the eastern bay, situated about 100 feet above the sea, was admirably suited for invalid residences; and he hoped that it would not be long before it was made available for the purpose. The Hôtels de la Paix, Grande Brétagne, de la Pension Anglaise, and the Grand Hotel, are in the eastern bay close to the sea, while at an elevation, away from the shore, is situated the Hôtel d'Italie. In the western bay, near the sea, are the Hotel Victoria—in which there is a lift, a great convenience for invalids—the Hôtel de la Méditerranée, and the Hôtel de Londres; outside the town, in the same direction, is the Hôtel Pavillon; while in more inland positions, away from the sea, are the Hôtels du Louvre and Beau Séjour; but there is, unfortunately,

bruna, a well-sheltered village, affording some accommodation, and situated 800 feet above the level of the sea, three miles from Mentone, on the Nice Road. My father, Dr. C. J. B. Williams, pointed it out to me as one of the warmest spots on the Riviera, during a tour which we made together through this region in 1856.

a lack of hotel accommodation in positions at a distance from the shore.

We have now reached the frontier, and in our next chapter we will consider the effects of the French southern climate on health and disease.

Its physical aspects have been examined, and it has been shown that the climate is warmer and drier than our own; that it has more sunshine, and counts fewer rainy days, and therefore gives greater opportunities for out-door exercise; and it is doubtless the principal recommendation of the south of France as a winter resort, that it affords facilities for out-of-door exercise in a pure, mild, and invigorating air, comparatively secure from the chills, changes, fogs, and wet or bleak weather, which pretty generally prevail in this country.

In reviewing the several localities in reference to their facilities for this health-giving exercise, Hyères and Cannes may be named as those which afford the greatest extent of sheltered rides, drives, and walks available for invalids. In the former place there are wanting only improvements in the roads and in the character of the vehicles to excel the other places in these respects. Cannes, with better roads, and a somewhat better supply of carriages and horses, does not enjoy an equal amount of shelter; but even in this point it surpasses Nice, which, with all the choice of equipage belonging to a large town, and with all the enchantment which beautiful scenery can give to its environs, has the disadvantage that in some of its beautiful drives the traveller may experience an excessive fall of temperature in the course of a few minutes. Mentone enjoys the protection of being almost hemmed in by mountains; but it thereby loses the advantage of numerous level walks and drives suitable for many invalids. It is, therefore, best adapted either to infirm patients who require a warm atmosphere with little locomotion, or to those much stronger and more active, who, on foot or on donkeys, can scramble up its mountain valleys

CHAPTER IV.

MEDICAL ASPECTS—EFFECTS OF THE CLIMATE ON HEALTH AND DISEASE.

The testimony of all the medical men practising among the natives is to the effect that, though acute disease occurs often, and is very fatal, chronic disease is rare; and many forms of degenerative diseases common in this country are scarcely known there. Dr. Griffith informs me, that no cases of Bright's kidney disease have occurred in the Hyères hospital within the memory of the present medical officers; and I may mention that Dr. Francis gives somewhat similar testimony concerning the hospitals of Spain: for he states that after carefully examining the principal hospitals of the Spanish peninsula for cases of this disease, he only succeeded in finding a few at Carthagena. I regret that I have been unable to collect statistics to decide the per-centage and mortality in each class of disease. Dr. Bennet states that the deaths from pulmonary consumption at Mentone are only one in every fifty-five of the total number, instead of one in five, as at Paris and London.* Mr. Richelmi,† who practised for 34 years in the Riviera, gives even stronger testimony. He states that out of 7000 deaths, which occurred at Villafranca, Monaco, Mentone, San Remo, and other places on the coast, only 107, or 1 in 65, were due to pulmonary consumption.

When we remember that the researches of Drs. Bowditch and Buchanan in America and England have, beyond doubt, established that *wetness* of soil is a cause of consumption to the population living on it, an explanation of the comparative exemption of this region from consumption is to be found in the great dryness of the soil.

Dr. Chambers, in his admirable Lectures on the Climate of Italy, has drawn attention to the registered mortality of Genoa, a city enjoying a climate much resembling that of the south of France, though liable to greater changes; and he has compared it with the registered mortality of London for the year 1862. Allowing for differences of nomenclature in

* One in eight is the mortality from phthisis, according to the Registrar-General's Report for 1862.'

† Thorowgood's Climatic Treatment of Consumption, p. 52.

the two registers, Dr. Chambers shows that, of the total number of deaths, the proportion from chronic disease is smaller at Genoa than at London—e.g. at Genoa it is 1 in 5·6, whereas at London it amounts to 1 in 3·2. He cites the following striking instances:—Anasarca, or general dropsy, caused 1 in 93 deaths at London; 1 in 239 at Genoa. Chronic affections of the respiratory organs, including asthma and bronchitis, but excluding pulmonary phthisis, caused at London 1 in 10 deaths; at Genoa 1 in 20. Aneurism at London caused 103 deaths in the year; at Genoa none. Chronic disease of the heart caused at London 1 in 27 deaths; at Genoa 1 in 33. No deaths from nephria, or kidney disease, were registered at Genoa, but probably some were included under deaths from anasarca.

Dr. Chambers remarks that the more decidedly chronic and degenerative the disease is, the more marked is the difference between the two cities. On the other hand, the registers show the proportion of deaths from acute disease (excluding zymotics) in Genoa to be more than double that of London—i.e. at Genoa 1 in 3·3; in London 1 in 7·7 deaths. Acute affections of the respiratory organs caused in Genoa 1 in 9 deaths; in London 1 in 16. Acute affections of the intestinal canal (including enteritis, gastritis, diarrhœa, and dysentery) caused in Genoa 1 in 8·9;

in London 1 in 30·3 deaths. Acute affections of the nervous centres caused in Genoa 1 in 59; in London 1 in 119 deaths. Apoplexy and cerebral congestion caused in Genoa 1 death in 12; in London 1 in 40. Acute inflammation of the heart caused in Genoa 1 death in 44; in London 1 in 606.

These comparisons are very striking; but it must be allowed that the force of them would be greater if we could be quite certain that the diagnosis of disease is as accurately carried out at Genoa as it is in London. The entire absence of deaths from aneurism seems almost improbable among an industrious population, a large number of whom are engaged in the laborious task of lading or unlading heavy ship cargoes, and in other occupations which exercise a decided strain on the vascular system. The high rate of mortality from acute disease of the respiratory organs is owing, probably, to the sudden atmospheric changes to which Genoa is peculiarly liable, on account of its unprotected position; and in this respect the health-resorts which we have been considering enjoy a great superiority over the Italian city.

The general effects of the climate of the south of France on patients may be divided into negative and positive.

F

Negative.—e. g. The avoidance of the exciting causes of so many diseases—namely, cold and damp. Many invalids, particularly those suffering from phthisis, asthma, emphysema, and chronic bronchitis, by simply avoiding catching fresh colds, prolong life and escape much suffering.

Positive.—e. g. The stimulating influence of the air, and the abundant out-door exercise which can be taken in this region. The functions of digestion and assimilation are improved, the standard of nutrition is raised, healthy tissue is formed, and morbid deposits are absorbed and eliminated. This stimulating character of the air is to be referred partly to the saline breezes coming from one of the saltest seas known, and partly to its dryness. The effect of this last quality on the skin is remarkable; for it might be thought that perspiration is promoted by a dry atmosphere more than by a moist one; but it is really found that an arid state of the skin checks the superficial circulation and secretion, and that a certain amount of moisture in the atmosphere considerably increases the amount of sensible perspiration. In the south of France it takes a great deal to make one perspire, as I have myself experienced. This bracing effect of the climate is well seen in the cessation of the nocturnal sweats of phthisical patients, who, after a time, only perspire like other people—viz. during exercise. This dryness sometimes amounts to a hurtful excess; and I have known patients suffering from dry bronchitis, who have been obliged to add to the moisture of their apartments by hanging damp sheets in them, or by the diffusion of steam from hot water. It is probably to the stimulating quality of the air that the want of sleep, so common among visitors, is due. Patients seldom sleep so soundly as in England, and often for only a few hours of the night. Many take a siesta in the day. But evil results seldom follow from this wakefulness; for the nervous system, as Dr. Chambers remarks, being in a healthier condition, seems to require less repose, and refreshes itself more rapidly. This stimulating quality, which is to be found to its greatest extent near the sea, does absolute harm to patients already suffering from an excited state of the nervous and vascular systems; as in cases of hyperæsthesia, cerebral erethism, gastritic dyspepsia, and of inflammatory and feverish affections generally. Such patients ought rather to avoid the Mediterranean region of France; or, if they should go there, they should choose the inland climates of Cimiez and Hyères in preference to those nearer the sea.

Some information with regard to giving a preference to some of these localities over others for certain forms of diseases may be acceptable, and I will now state the results of my enquiries and experience on this point. It has been my object to show that the climate of the individual health-resorts depends on their proximity to, or their distance from, the sea; that the air of the places immediately on the coast is exciting, while that of the more inland is less stimulating and softer; and it is on these differences that the selection of a locality should be based. In cases of bronchitis the type must be borne in mind. If it be humid, accompanied by free expectoration, and devoid of febrile symptoms, Cannes,* Mentone, or even Nice, would be suitable. If it be dry, and attended by inflammatory symptoms, the softer climates of Hyères and Cimiez would be preferable. As regards spasmodic asthma, Dr. Bennet says that some cases do well at Mentone, but the majority do not derive benefit. This may be accounted for by the closeness of the

---

* Cannes answers particularly well when the amount of expectoration is in excess of the patient's strength, as in bronchorrhœa of the aged. Dr. Whiteley, of Cannes, informs me that the climate is very effective in stopping any discharge of the nature of a flux, and arising from a relaxed state of system, unaccompanied by inflammatory symptoms.

Mentonian atmosphere. Cannes does not always suit this disease according to Dr. Battersby; but the climate of Hyères has answered very well for many such cases. Cases of emphysema, in which the avoidance of bronchitis is necessary, do well at any of the localities except Nice. In pulmonary phthisis, as in bronchitis, the type must be taken into consideration.

Cases of non-inflammatory phthisis in all stages, except the last, derive marvellous benefit from the climate of this region. For these Cannes, Mentone, Hyères, and the suburbs of Nice are suitable; Nice itself, on account of its sudden changes of temperature, is less safe.

The inflammatory form, accompanied by fever and gastric irritation, is better at Hyères and Cimiez. Dr. Battersby does not recommend Cannes for such cases. According to Dr. Siordet, the effect of the Mentonian climate in these cases is to bring on hectic fever and to hasten the fatal termination.

The Mediterranean climate generally is, I think, too dry and stimulating for this form; and a moister, softer atmosphere is desirable. Such can be found, although at a sacrifice of warmth, in some of the British watering-places, e. g. Torquay, Ventnor, Bournemouth, &c., where saline breezes are combined with a lower mean temperature and a considerably greater amount of moisture. Should the large number of wet days, and the prevalence of wind in these localities during the winter months, present objections, the dryer and remarkably still atmosphere of Pau may prove more beneficial. It may be well to direct our attention for a short space to this interesting climate.

Pau is eighteen hours distant by rail from Paris, and stands on a terrace of gravel at an elevation of 150 feet above the river. It faces south towards the Pyrenees, distant about twenty miles, and overlooks a rich valley through which the clear Gave flows. According to Sir Alexander Taylor, 'it is protected on the north by the Landes of Pont Long, which ascend very gradually to the distance of eight miles from Pau. The north wind is thus directed into currents, which being attracted by the lofty mountains to the south, pass at an elevation considerably above the town: so that the clouds may be often seen quickly sailing onwards, when the leaves are unmoved on the lower level.' The town is to a certain extent screened from the north-west and west winds, but it is open to the southerly and easterly breezes. The chief feature of the climate is the wonderful stillness of the atmosphere. The leaves of the trees scarcely move, and the rain descends almost perpendicularly; and even when the wind does blow, its gusts are of short duration, and of no great force. Sir A. Taylor states that during a period of six months' observations the presence of wind was recorded only 13 times. The mean winter temperature is 42·8° Fahr., 5° Fahr. lower than at Nice; and snow falls occasionally. The average rainfall is 43 inches compared with 25 at Nice; and the average number of rainy days is 119, compared with 60 at the latter place. My friend Dr. Bagnall has kindly forwarded the hygrometrical observations for the last two winters, and, on comparing them with those of Nice and the other Mediterranean resorts, I find that during the winter 1865-66 the average difference between the bulbs amounted to 2·7° Fahr. at Nice, and to 2·4° Fahr. at Pau, the Kew difference for the same period being 1·65° Fahr.

The hygrometrical results for the winter 1866-67 at Nice and Pau tallied with those of the preceding year, but the month of March, 1867, presented a marked contrast at the two places.

At Nice the average difference between the bulbs was 2·3° Fahr., while at Pau it was 4·2° Fahr.; so that during this month Pau actually enjoyed a drier climate than Nice. This was probably exceptional.

When we consider the large amount of rain that falls at Pau, and the prevalence of moist westerly breezes, it is rather surprising that the hygrometer should not indicate a greater amount of humidity during the winter, and the phenomenon can only be accounted for by the absorbent nature of the gravelly soil on which the town stands, and the natural 'draining' of the place by the Gave stream. The calmness of the atmosphere, combined with a certain degree of moisture, imparts a sedative quality to the climate, which acts beneficially on the inflammatory type of phthisis, and Pau can be recommended for this form of disease. The number of rainy days, and the decidedly cold and cloudy weather to which Pau is at times subject, prevent the same freedom of exercise which can be taken in the more southern region, and which is of great importance to a large number of pulmonary invalids; but in fine weather Pau has the recommendation of a large choice of beautiful drives, which cannot be said of all the Mediterranean resorts.

In our brief notice of the claims of different localities as a winter residence for pulmonary invalids, we must not omit the mention of Madeira, as one which was long supposed to stand foremost in the list. In the last few years this island has not maintained its place in the favour of the profession or of the public; and its comparative desertion has been ascribed by a recent writer to its 'seeming rather to satisfy the requirements of the bygone period of the professional mind, when pulmonary consumption was considered a species of inflammatory disease, than to satisfy present requirements,' when 'phthisis is considered a disease of debility, of anæmia, of organic exhaustion, and of defective nutrition.' Now, although we may admit that the advance of pathology has thrown light on both the nature and treatment of consumptive diseases, we can hardly allow that there is such a change in the 'professional mind' as to deny altogether to inflammation any share in producing or aggravating these maladies. It is still the opinion of those who have had the largest experience, that a considerable number of cases of phthisis take their origin from inflammatory attacks; that intercurrent inflammations are the most common causes of acceleration of the disease in this country; and that a climate which supplies fresh air without cold, damp, and sudden changes, owes much of its salutary influence to its excluding these causes of inflammation. The declining popularity of Madeira may be attributed partly to its distance from England,

and its isolation, involving the necessity of an irksome sea voyage, and removing the patients so far from friends and home; and partly to a temporary cause—the establishment of a quarantine—which, in latter years, proved a most serious obstacle to the reception of invalids. From recent information it must be added, that the climate of Madeira is deteriorated in its salubrity, in consequence of the substitution of the sugar-cane for the vine, since the ravages of the oïdium have destroyed the productiveness of the latter. In the cultivation of sugar-cane constant irrigation is required, which keeps the ground in a damp state; and the evil effects have already become manifest in the prevalence and mortality of fever in the island. Even before this change took place the air of Funchal was never healthy; and for many years my father's experience led him to the conclusion that patients rarely benefited much by wintering in Madeira, unless they were strong enough to ride daily on the mountain roads above the town.

The climate of Funchal and the neighbouring country is remarkable for its steady mildness throughout the winter, being exempt from sudden changes, and from wind, except to a very limited extent; and it counts few rainy days: but the air contains a considerable quantity of aqueous vapour, which imparts to it a softness, highly beneficial in the inflammatory type of consumption, but relaxing and enervating for other forms, and even for healthy persons. It may therefore be said generally, that Madeira is best suited for the early stages and inflammatory forms of the disease, and is not likely to benefit those in whom the cachexia or degenerative tendency is well marked.

It may not be out of place to notice the experiment which was lately made, by the Committee of the Hospital for Consumption at Brompton, in sending twenty patients, selected for the purpose, to Madeira, for the benefit of the climate in winter. The result was by no means so favourable as had been anticipated: not more than three out of the number returned in an improved condition. One died in the island; and the others generally lost flesh and strength; and the disease in them made considerable progress. .It must be added that these patients, although well fed and cared for, had little or no medicine administered to them, and they thus show that the climate alone, unaided by cod-liver oil, tonics, and other aids in treatment, is truly of inferior efficacy.

CHAPTER V.

HYGIENICS OF CONSUMPTION—FOOD—CLOTHING—VENTILATION—EXERCISE.

A few remarks on the hygienics of consumptive patients will not be out of place; comprising hints with respect to food, clothing, ventilation, and exercise in this region; as these may aid the beneficial influence of the climate, and considerably increase the probability of the patient's recovery.

Food.—As might be expected in a country devoted to the culture of the olive, vine, orange, and lemon, there is a scarcity of pasture for sheep and cattle; and consequently meat is neither so abundant nor so good as in the grazing countries of the north. It often has to be brought some distance, and is seldom really tender. Added to this, the cooking, except in the best hotels, is not of that plain wholesome character which best agrees with the delicate appetite of an invalid; and often contains a large amount of grease and oil, which in a warm climate tends to produce biliousness and liver disturbance. By strict injunctions to the cooks, this objectionable quality in the food may be corrected, and a tolerably wholesome diet procured, and it is of the utmost importance that this point should receive due consideration; for not only may inattention to it induce bilious derangements, but it may prevent the regular and sufficient administration of cod-liver oil, which is the most potent of all means to promote recovery in phthisis. It is mainly through want of attention to this point, that an impression prevails with many practitioners in the south, that patients cannot take the oil in warm climates. But we now have the best evidence, that not only in the south of France and Italy, but in Madeira also, and even in India, this remedy can be given regularly and freely with excellent results. A patient will find that the two daily doses of the oil, with the butter eaten at meals, will be as much, in the way of fatty matter, as his digestive powers can manage in this region, and any extra supply in the form of rich dishes will only disagree. If he avoids these, and only remembers to take the oil *immediately after* meals, he will be able to continue its use through the winter, and will gain flesh and strength thereby. The bread is tolerable, and the butter excellent; but the latter is not the produce of the country, being imported from Milan and other distant places.

The vegetables of this region are abundant and in perfection. Their variety, their profusion, and the rapid succession of their crops in winter cause considerable surprise to visitors. They are astonished to find themselves eating new potatoes in January, and deliciously tender asparagus in March; while all through the winter they can depend on an abundant and constant supply of green peas, dwarf beans, cauliflowers, brocoli, and artichokes, all excellent in flavour and quality. The artichokes are particularly tender, and the cauliflowers of very fine growth. But it is in the form of fruit that we see the effect of that bright sun, those clear skies, and that genial soil, most marked. Nature, aided but little by Art, pours forth a bounteous feast of oranges of various kinds and sizes, of lemons sweet and acid, of shaddocks, pomegranates, almonds, apples, and pears.

The Japanese medlar, a tree very sensitive to the slightest degree of frost, bears a delicious and highly prized fruit; and dates are produced from the palms, but it must be confessed they are hardly eatable. The oranges of Nice are particularly refreshing to the invalid in this warm climate, and the fragrant odour of their rind is only equalled by the sweetness of their taste. These and many other fruits the invalid can enjoy throughout the winter; and provided that he does so in moderation, he will suffer no injurious effect from them.

With regard to alcoholic beverages, it may be stated that the stimulating influence of the climate considerably diminishes the necessity for their use. Spirits and the stronger wines, port and sherry, are quite out of place. As Dr. Chambers justly observes, 'one does not feel to want them. A single glass of Orvieto or Capri there seems to produce as much exhilarating relief as an allowance of the domestic port and sherry containing five times the quantity of spirit.' The same influence is to be taken into account in prescribing medicines. The strong tonics are as a rule to be avoided, and the milder ones substituted.

Clothing.—This region owes much of the warmth of its winter climate to the prevalence of sunshine; in which one may bask with all the sensations of a fine English summer: but cloudy days and even shady places have their chilly influences, against which the patient should be protected by always wearing a flannel or merino underdress. This is especially required in driving in an open carriage: and the more so, early in the morning, or late in the afternoon, when invalids should be provided with a shawl or overcoat. Light-coloured suits of thick texture are the most suitable; as, while sufficiently warm to prevent the differences of sun and shade being felt, they do not absorb the sun's rays so much, nor present so sorry a figure under the influence of dust, as the darker ones. The light-coloured parasol with a dark lining is useful to protect the head from the sun's heat, and the eyes from its glare; and is freely used not only by ladies, but also by gentlemen, who are wise enough not to allow the effeminacy of the proceeding to prevent their experiencing its comfort. The great object to be aimed at in clothing is to render the body independent of slight changes in the temperature resulting from alternations of sun and shade, and at the same time to protect the head from the evil effects of powerful sunshine; for in this climate sunstroke is far from uncommon, although it is easily guarded against by the above precautions.

Ventilation.—Whilst we fully admit the importance to a consumptive patient of pure air and its abundant supply, to improve the assimilative and nutritive powers, to raise the standard of products, to facilitate the formation of healthy fibrin and albumen, and to prevent the production of degraded tissues, as tubercle;

in short to remove the tuberculous cachexia,—we must not overlook certain conditions necessary to ensure its full and beneficial influence. The pure air must be of a certain temperature, or at any rate within a certain range, otherwise the alternations of heat and cold will induce fresh attacks of catarrh, bronchitis, pulmonary congestion, and possibly pneumonia, any of which complications may considerably aggravate the original disease. The practice of sleeping with open windows, sometimes recommended to consumptive patients, is a very questionable one. It is questionable during summer in England, where there is less variation of temperature, owing to the influence of the large quantity of aqueous vapour present in the atmosphere, which reflects back to the earth the heat lost by radiation, and thus prevents a great fall of temperature. Even with this equalising influence of the British climate, the coolness and dampness of the night air in summer is often sufficient to induce chill of the body and its sundry evil consequences. But in the south of France the practice is doubly hazardous; for here, during the winter, the fall after sunset is rapid, and the diurnal range very considerable; owing, as I explained in the commencement of this work, to the amount of moisture present in the atmosphere being small and the effects of radiation in a clear sky being unchecked. The temperature may be that of summer at noonday, and towards midnight it may fall to 40° Fahr.; and it is this change that is a fertile cause of bronchitis, pleurisy, and pneumonia in consumptive cases. Apart from the general chill of the body, which may sometimes be obviated by an abundant supply of blankets and coverings, the local effects on the lungs produced by the changes in the air are of some importance. Against these nature has to a certain extent provided a safeguard in the nasal passages, where a large and tortuous plexus of veins furnishes an apparatus for warming the air previous to its entry into the lungs. The mouth, however, is not equally well armed, and air entering by that passage is not so likely to be raised to the proper temperature for respiration as the portion which passes through the nose. Many persons sleep with their mouths open, and breathe almost entirely through them, and in these the inspired air loses the warming influence of the nasal passages. How common it is for even healthy persons, after sleeping with their bedroom windows open, to awake the next morning with a slight cold in the head. This, in the case of the strong, passes off in the course of a few hours, but in pulmonary invalids may lead to further mischief.

One of the great objects in the treatment of phthisis being the avoidance of all secondary complications, it is evident that the patient should be protected from this injurious influence as much as possible, and it is better for him to be kept too warm than too cold. He will have full opportunities in the daytime to enjoy a pure and warm atmosphere, and if the air of his bedroom at night be not so fresh and pure as that which he has been breathing in the day, he can console himself that it is far better to inhale it, than to subject his frame, especially in a dormant state, to all the variations of night temperature. After all, it must never be forgotten that cold is a greater evil than heat, especially in lung complaints; and it is wiser therefore, after airing the bedroom with widely opened windows in the day, to close them at nightfall and to trust to the chimney and to an occasional fire for ventilation.

Exercise.—It may be broadly stated that, in all cases of phthisis, exercise in some form or other is beneficial; but whether it should be of the active or passive kind, and what varieties of each are admissible, depends on the stage and type of the disease, and also on the strength of the patient. In the early stages, where the symptoms are not active, where there has been no recent blood-spitting, and where the cough is not hard or frequent, those varieties of active exercise are of most advantage which most effectually expand the upper portions of the chest, thereby bringing into play the upper lobes of the lungs, so generally the seat of tubercle; and by causing the blood to circulate freely through the pulmonary tissue, thus prevent local congestions and fresh deposits, and aid materially in the absorption of old ones.

What are the varieties of exercise which best accomplish this end? Those in which the upper extremities are raised, and the muscles connecting them with the thorax brought into activity. When the arm is raised, the numerous muscles which arise from the ribs, and are inserted into the bones of the upper extremity, e.g. the pectoralis major and minor, the subclavius, the serratus magnus, &c., in contracting, raise the upper ribs, and thus increase the size of the chest cavity. This necessitates the inspiration of a larger amount of air. Dr. Silvester has called attention to this important principle, and on it has founded his excellent system of restoring respiration in cases of drowning, narcotism, &c. He has also recommended a modification of it in the incipient stages of phthisis. The forms of exercise which carry out this principle are—rowing, particularly the pull and backward movement; the use of the alpenstock in mountain ascents; swinging by the arms from a horizontal bar, or from a trapeze; climbing ladders or trees. Dumb-bells, as commonly used, are calculated to develop the arms more than the chest; and rather tend to depress the latter, by their weight. Various special gymnastic exercises, of which there is a great choice and variety nowadays, may more or less answer the purpose; but there is one form which is particularly applicable to the object above mentioned, viz. the *gymnast*, invented by Mr. Hodges, 89, Southampton Row. But to make this instrument answer the purpose of a chest elevator or expander, it should be fixed, not as usually done, at the height of the operator, but considerably above his head, in or near the ceiling, with the handles reaching down about to his shoulders; then by holding the handles, and walking a few paces forwards and backwards, the arms are brought into a species of action, which, while it exercises the whole body, especially tends to expand and elevate the upper part of the chest.

Walking exercise, as a rule, does not work the upper extremities, or raise the upper ribs, but acts generally on the system, by drawing the blood to the extremities, and quickening the circulation through the lungs. In mountain ascents, and in fast walking, the quickening of the circulation brings the whole lungs into play; and in this way the upper lobes come into full use. If the alpenstock be used in mountain climbing, the beneficial local effects of raising the upper ribs may be combined with the general advantages of walking.

Walking exercise can be taken in all stages of phthisis, provided there be no active symptoms present. Even where cavities are formed, if there be no recent inflammation, a limited amount, and performed on level ground, is beneficial; but great care must be taken not to overtax the patient's strength.

Passive exercise may be used by the weak and delicate, even in advanced stages of phthisis, or when it is of the inflammatory type. Open carriage exercise, sailing, or being rowed in a boat, are instances, in all of which little muscular exercise is involved; and they may be considered as means of supplying a constant change of air, with the least fatigue, while their effect in improving the circulation and appetite, and in promoting sleep, is often very apparent. But even these make some demand on muscular and nervous power, and must not be carried to the extent of producing exhaustion in weak subjects. The roughness of many of the roads, and the not always luxurious construction of the carriages in the south, sometimes invalidate the distinction between active and passive exercise. Riding may be considered intermediate between the two forms, for rapid paces, and even the management of a high-mettled steed, may well be classed as active exercise; but the general mode of equitation in the south, on donkeys, mules, or very quiet ponies, hardly rises above the standard of passive exercise.

From the time of Sydenham, horse exercise has always been acknowledged to be peculiarly beneficial to those pulmonary patients that can bear it; and any one who closely studies the effect of the motions of riding, on the respiration, may perceive that there is a remarkable relation between the process of breathing, and the paces of the horse. This, in addition to the inspiriting effect of the air and exercise, may have its share in the salutary results.

CHAPTER VI.

WINTER STATIONS OF ITALY—BORDIGHERA—SAN REMO —NERVI—PISA—ROME—CAPRI—SALERNO—AMALFI.

IF after quitting Mentone we cross the Pont St. Louis, and enter Italy, we shall find along the beautiful Riviera route several localities, sheltered from cold winds by the Maritime Alps, and enjoying the full effects of the sun's rays and the Mediterranean sea.

These localities, though not as yet, with the exception of San Remo, equal in point of accommodation to the French health resorts, are rapidly developing their resources; and with regard to position, have already proved to be safe wintering places for many pulmonary invalids. As it is advisable that this class should have a large selection of southern refuges from northern winters, I propose in this chapter to mention briefly some of the more sheltered spots of Italy; regretting much that, through lack of meteorological data, I cannot speak with certainty as to the climate in every case.

Bordighera lies about eleven miles from Mentone, on a promontory jutting far out into the sea, from which exquisite views are obtained of the coast as far west as the Estrelles. The splendid palm-groves, descending even to the shore, and containing trees upwards of 1000 years old, form the chief feature of the scene, and testify powerfully to the warmth of the climate. According to the observations of the Rev. A. Craig, the English chaplain, for which I am indebted to Dr. Daubeny, of San Remo, the mean temperature, during the months of January, February, and March, 1867, was lower than that of San Remo; but the daily range was less. The hygrometrical observations taken at 9 a.m. showed a difference of 2·8° Fahr. between the bulbs, as compared with 2·4° Fahr. at Nice for the same period.

The Hôtel d'Angleterre is good; and invalids have passed the winter in it with tolerable comfort to themselves, and benefit to their health.

San Remo, a town of 11,000 inhabitants, about three and a half hours' drive from Mentone, rises like a white pyramid from the shore of the Mediterranean; contrasting strongly with the sombre colour of the vast olive-grove, which extends for miles in its neighbourhood, covering the nearer range of hills, forming the principal feature of the scenery, and contributing, by its produce, in no small degree to the prosperity of the town. The protecting range, sloping gently back from the shore, is intersected, but not penetrated, by numerous valleys remarkable for the luxuriance of their vegetation, and running out on either side to the sea, forms the Capes Nero and Verde, which enclose a fine bay four miles in width. The nearer and olive-clad chain is backed by one more lofty, which rises in the Monte Bignone to the height of 4270 feet, and on which the vegetation is less abundant and more hardy, including varieties of the pine and oak. The town itself faces south, and is built on a spur of the nearer range, with a valley on either side. It is well sheltered to the north, north-east, north-west, and east; but it is exposed to all the southerly winds, including the south-east.

Lemon-trees abound, and, from their thriving state and the size of their fruit, show how rarely frost occurs. Palms are to be seen, and the olive oil produced is of excellent quality. Villas are sprinkled over the hillsides in beautifully sheltered positions, many of them at a distance from the sea; and some of the hotels are well placed; being outside the town, and removed from the disagreeables of noise and bad drainage.

On the western side of the town is the Hôtel de Londres, and on the eastern the Hôtels Victoria and d'Angleterre, while in the town itself is the Hôtel d'Italie.

The mean temperature is 49·1° Fahr. (Sigmund): nearly as high as Dr. Bennet's estimate of that of Mentone, and it would appear from a comparison of the thermometrical tables kept by Dr. Daubeny, one of the resident English physicians, with those of Dr. Bennet for the same winter (1864–65) at Mentone, that the mean range of temperature is much less at San Remo. At Mentone it amounted to 12·6° Fahr.; at San Remo to 9·7° Fahr.: leaving a difference in favour of the latter place of nearly 3° Fahr. On comparing Professor Goiran's record of mean temperature for the months January, February, and March, 1865, with that of Nice, I found a difference of nearly 2° Fahr. in favour of San Remo. The number of rainy days is 45; the smallest number of all the places described; and Dr. Daubeny found the average difference at 9 a.m. between the bulbs, for the winter 1866–7, to amount to 3·7° Fahr.: while at Nice it was 2·7° Fahr., and at Hyères 2·9° Fahr.

The climate is warm and dry; and from the protecting ranges not rising precipitously, as at Mentone, but sloping gradually back, the shelter from northerly winds is not quite so perfect as at the last-named place. At the same time the vast olive-grove screens the locality from any cold blasts, and the breezes which filter through the olives impart a pleasing freshness to the atmosphere, and remove sensations of lassitude often experienced in well protected spots. The size of the sheltered area gives patients a considerable choice of residences, which can be found either close to, or at varying distances from, the sea, according to the requirements of the case: while the numerous wooded valleys, abounding in exquisite wild flowers, provide plenty of donkey and foot excursions.*

As far as San Remo the resources of this beautiful and sheltered coast-line may be said to have been fairly developed, and an invalid can find good accommodation at any of the places we have described. Beyond San Remo, on the Genoa road, and again between Genoa and Spezia, the case is different: sheltered places suitable for winter residences are not wanting; but unfortunately, with some few exceptions, the accommodation is not sufficiently good for invalids. Considering the large and increasing number of visitors, and the crowded state of some of the

---

* For further information on San Remo I would refer the reader to Mr. Aspinall's little work, entitled 'San Remo as a Winter Residence, by an Invalid.'

French health-resorts, it is of the greatest importance that fresh ones should be established on the Italian side, and I will therefore notice some spots which, from their sheltered position, appeared to me, during a recent visit, worthy of being chosen for winter stations. After passing Oneglia we come to the spacious valley of *Diano*, open to the sea, but defended from the cold blasts of the north, north-east, and north-west winds by a fine range of hills, clothed with olives, 'prodigal in oil and hoary to the wind.'

Several wooded dells run back towards these hills without penetrating them, and, what is of some importance, but few torrents traverse the valley. It contains some villages, rather devoid of accommodation; *Diano Marino* on the shore, inhabited chiefly by fishermen, and *Diano Castello*, which peeps out, half-hidden, from the olive-covered slopes above. The former might supply the nucleus of a marine residence; and the latter that of an inland one. *Alassio*, about 14 miles from Oneglia, lies on the sea margin, flanked like Diano by olive-clad hills, and seemingly enjoying much shelter. On the eastern Riviera, between Genoa and Spezzia, there are several favoured spots, whose resources the completion of the railroad will probably develope, and I may mention that, during my last visit to this portion of the coast, the mistral was blowing with great violence, and I thus had a good opportunity of judging of the amount of protection enjoyed by each locality.

Nervi, about an hour's drive from Genoa, stands on a ledge of level ground, which descends in rocky masses into the Mediterranean; here affording boating, bathing, and tolerable fishing. The place is well sheltered to the north and east by wooded hills, on whose sunny slopes the orange and lemon flourish. Olive groves, besprinkled with villas and picturesque hamlets, rise about the village; and the view from this spot of the Bay of Genoa and the Col de Tende range is exceedingly beautiful. A good hotel and moderate boarding house (Hôtel de la Pension Anglaise) with well sheltered grounds, pleasant walks and rides, and its short distance from Genoa, render Nervi attractive to invalids, many of whom have safely wintered there.

A magnificent view of the Gulf of Genoa is obtained by ascending the bold promontory of Ruta, the rocky coast of the Western Riviera, above which towers the snowy range of the Col de Tende, contrasting with the gentler, and yet scarcely less exquisite features of the Eastern Riviera, where a succession of hills, rich in varied vegetation, slope, curve within curve, and outline within outline, in voluptuous beauty to the sea; the scene enlivened by picturesque hamlets above, each crowned with its pretty campanile, while below are the busy little ports, gay with the many-coloured sails of the feluccas which glide so gracefully over the blue expanse of the Mediterranean.

The northern side of the Ruta promontory is, from its aspect, unfit for a winter residence; but the southern is beautifully sheltered from all northerly winds, and abounds in luxuriant vegetation, extending even to the water's edge. On the coast, facing south-east, is a good-sized village called *Santa Margherita*, near which appear to be several desirable sites for building; but, curious to say, no carriage-road connects the place with the mainland; communication being kept up by water only.

Passing Rapallo, where the mistral was only too plainly felt, the road ascends again, and we enter a lovely bay, on whose slopes, clad with olives and vines, nestles the pretty village of *Zoagi*, containing some villas. A few miles further is *Chiavari*, the west end of which is charmingly placed on the southern side of a bold ascent, but the east end is slightly exposed to the mistral; here there is tolerable hotel accommodation. *Sestri di Levante* has been mentioned as a future wintering place, but no one who has visited it, when the mistral is blowing, can fail to agree with Dr. Bennet, that it is quite unfit for the purpose.

The Italian Admiralty Works preclude *Spezia* from becoming a winter resort; though its neighbourhood, as furnishing sites for invalid residences, fully deserves the encomium passed on it by Dr. Chambers. Leaving the Riviera, we now come to—

Pisa, a most interesting city, situated six miles from the Mediterranean, on the banks of the swift-flowing Arno. It lies in the midst of a large plain, parts of which are marshy; and as far as I could judge, after inspecting its position from the top of its far-famed leaning tower, it does not enjoy much protection on any side.

The Apennines, which rise to the north of the town, and are snow-capped in winter, are too distant to afford it much shelter. On the east the Tuscan hills give some protection; but it is open to winds from all the other quarters. The mean winter temperature is $44·8°$ Fahr., and the annual rainfall 45 inches. The climate is moist, unstimulating, and sedative, and bears certain resemblances to that of Pau: but the locality is damper and not so well sheltered. The Lung' Arno, the principal promenade, is tolerably well protected: but excursions in the neighbourhood, particularly the one to 'Le Cascine,' are open to the dangers which arise from exposure to cold winds. The climate of Pisa has a high repute among the Italians themselves; and its sedative character may perhaps recommend it for the inflammatory type of phthisis. The hotel accommodation is excellent, the Hôtel de la Grande Bretagne being one of the most comfortable and moderate in Italy.

Rome was formerly recommended by Sir James Clark and other eminent physicians as a winter residence for patients in the early stages of phthisis, and for some other forms of pulmonary disease.

It enjoys a higher mean temperature in winter than might be expected from its isolated position in the midst of a vast plain, and situated at a considerable distance from the surrounding mountains. The range of temperature is great; and the number of rainy days rather large when compared with the numbers at Nice and in the Riviera. It is exposed to the influence of winds from all quarters, and notably of the *Tramontana*, or north wind; and the *Sirocco*, or south-east wind; and the alternations of temperature accompanying the changes between northerly and southerly winds render Rome by no means a safe winter residence for pulmonary invalids. The surrounding campagna, with its marshes, exercises an undoubtedly malarious influence on the city and its neighbourhood; and unfortunately visitors are liable even in winter to suffer in some way or other from this influence. Intermittent and remittent fever are the most severe forms of disease; but to the same cause may be traced the headaches and *migraine* often complained of at Rome.

My colleague, Dr. Pollock, who practised for seven years at Rome, informs me that during that period he always had cases of intermittent or remittent fever under his care, even in winter. The sources of this fever, its connection with the depopulation of the country, the extent of its influence, and the conditions which modify its prevalence, form topics too comprehensive to be treated of here; but as, during a recent visit, I paid considerable attention to the subject, a passing notice of the principal causes of the fever, with a view to their prevention, may prove of service to some of my readers.

The malaria seems to have a near origin in the neighbouring campagna; but it has a more remote one in the Pontine Marshes, a very dangerous district about thirty miles south of Rome; for in the summer months, when the wind blows from that quarter, an increase of fever takes place, especially at the end of the city lying nearest the marshes.

Again, some parts of the city itself appear to be sources of malaria, while others enjoy perfect immunity, and it is worth noting that the more airy and fashionable quarter of the town, which includes several open spaces, as the Piazza di Spagna and the Piazza del Popolo, is generally the first attacked; while the crowded dirty Ghetto, or Jews' quarter, with its narrow streets, bad drainage, and deficient ventilation, often escapes altogether. One of the best ascertained facts with regard to Roman malaria, is that it diminishes directly as the population increases, and *vice versâ*, that it increases as the population diminishes; and in this way is explained its great spread since the days of the empire; for the population of the city has fallen from upwards of 2,000,000 (Dr. Dyer's * estimate) to 180,000, the present number of inhabitants; and the campagna, once thickly peopled and well wooded, is now bare, and contains but a few scattered towns and villages. The chief agents involved in the production of this malaria appear to be,—firstly, the sun's rays acting on the moist soil, and thus promoting evaporation. Fever is found to prevail in proportion to the sun's power, and the

* Article 'Roma,' Smith's 'Greek and Roman Geography.'

states of the atmosphere, which determine the amount of sunshine, exert no small influence over the malaria. When the sky becomes cloudy, or when the rain falls heavily,—when, in fact, the conditions necessary for evaporation cease for a time, a diminution in the number of fever cases takes place. Secondly, the soil, which in the malarious districts generally consists of pozzuolana, a species of tufa, is porous and has a remarkable power of absorbing moisture. A piece of it when brought into the house becomes quite damp, and in a short time will be found to have increased in weight. In the campagna, when the foundation of a house is dug, it is often necessary to have a pump continually at work to prevent the accumulation of water. The sun's rays striking on so moist a soil causes it to exhale quickly, and rapid evaporation, the condition necessary for the production of the poison, is the result. We find, too, that obstacles to evaporation are also obstacles to the production of malaria; as the obscuring of the sun's rays by cloud on the one hand, and the covering of the soil by buildings and pavement on the other, both tend to arrest the development of the miasma. In this way we can explain the unequal distribution of the fever in Rome. The Ghetto is closely built over, and well paved, and thus the action of the sun on the soil is virtually prevented; whereas the Piazza del Popolo is not covered with pavement, and therefore offers no obstacle to evaporation.

My friend Dr. Topham, one of the resident physicians, suggests an ingenious explanation of the malaria, founded on the water system of ancient Rome. To obtain a sufficient supply for the vast population, whose bathing propensities are shown by the colossal baths of Titus, Caracalla, and Diocletian, water was conveyed to Rome by fourteen magnificent aqueducts, and these, some of them reaching between 40 and 50 miles, conducted the streams of the surrounding mountains into the heart of the city. Thus a large supply of water was collected, and now that only three aqueducts are used, and the rest broken down and in ruins, this body of water, diverted from its original course, spreads itself over the campagna, rendering it marshy and deleterious.

The prevalence of acute disease of the respiratory organs at Rome, to which so many physicians bear testimony, may be ascribed to the vicissitudes of temperature, and not to any stimulating influence of the climate, which, on the contrary, is rather moist and sedative, and somewhat resembles that of Pisa. The air is soft and warm in fine sunny weather; and when the Tramontana and Sirocco are not blowing, there is a calmness of atmosphere which is suitable to some forms of chest disease. But, according to my experience, few places are so unfit for a pulmonary invalid as Rome. If he wanders on the sunny and sheltered slopes of the Pincian he is tolerably safe; but a drive into the campagna exposes him to cold winds, and possibly to malaria; a walk in the well-like streets, so narrow and so walled in by lofty palaces that the sun hardly reaches the pavement, is not beneficial; and visits to the icy museums and galleries, to which his inclinations and his friends direct him, are often fraught with grave dangers.

I may, however, generally state, that a careful survey of the large cities of Italy, such as Rome, Florence, and Naples, has convinced me, as it has many other medical men, that the excitement and fatigue of sight-seeing, the gaiety, and the exposure to extreme temperatures, which a residence in them commonly involves, render them quite unfit winter stations for pulmonary invalids. These, in the midst of the temptations offered by the treasure-houses of ancient and modern art, forget the objects for which they have been exiled from home, and foolishly expose themselves to cold, fatigue, and other dangers: thereby frustrating the intentions of their physicians, who have recommended a winter in the south; relying that patients, if they quit their native land for the improvement of health, will have the good sense to keep the object of their exile always in view, and not expose themselves to unnecessary and often fatal dangers; but rather devote their attention wholly and solely to reaping the full advantages to be derived from the substitution of a southern for a northern winter. Far safer and wiser is it to lead a quiet and tranquil life in one of the above-described health resorts, apart from the fascinations of draughty picture galleries, and cold museums of sculpture; and to rest contented with the ever-varying and beauteous feast of nature's charms, bounteously spread and safely partaken of, and the enjoyment of which is doubly enhanced by a knowledge of the great laws of natural science, which by directing the attention to the wonderful harmony of nature, incites us to show forth the same in our own lives.

Besides the above objections to the large towns of Italy as winter stations, Naples lies under a still more powerful one, in its bad drainage; which is so arranged that a large amount of the sewage of this densely populated city is conducted through the visitors' quarter, and discharged on the shore, or into the almost tideless sea, in front of the public gardens and principal hotels. The Chiaja, where the chief accommodation for strangers is to be found, adjoins this outfall of large sewers, which directly communicate with the external air by means of holes constructed for the purpose of carrying off the rainwater, but serving also as vent-holes to conduct the smells and noxious vapours into the thoroughfare and the neighbouring houses. Should a pulmonary invalid wish to escape from the dangers of Naples, arising from its exposed situation and its unhygienic state, he may find refuges in its neighbourhood, not equal in point of protection to the French stations, but still having a certain amount of shelter from cold winds.

The beautiful island of Capri, whose rocky outline rises in full view of Naples, and from the brilliant transparency of the southern atmosphere often appears but a short distance from the shore, enjoys a mild and equable climate throughout the winter.

A well-known English resident, who has much benefited by the climate, writes:—' The formation of the island is limestone, and it is therefore very dry. Very little rain falls there, or much less than on the mainland, as the clouds are carried through the channels on either side; and the peasantry often look with longing eyes on the rain, which is falling on the continent. The southern part of the island is perfectly sheltered from every breath of the north wind, and here the vegetation is more luxuriant and earlier than in the northern portion.'

The town of Capri, and several of the villas, are protected from most of the northerly winds by the precipitous rocks of Anacapri; but there is no shelter from other winds, and the island is occasionally subject to stormy weather. Wherever there is sufficient soil to till, the vegetation is rich, as shown by its fruits, oil, and far-famed wine; and a variety of the fan palm is to be seen growing wild in the crevices of the rocks on the north side of the island. Capri is nineteen miles from Naples, and can also be reached by boat from Sorrento. The hotel accommodation is good; and patients have passed very tolerable winters there; but communication with the mainland is not always open.

Salerno is situated on the northern shore of its gulf, at the foot of the St. Angelo mountains, which screen it from northerly and easterly winds, and afford it greater protection than is enjoyed by any town on the Gulf of Naples. It faces south; and the force of the Sirocco wind is broken by the Posidian promontory, which limits the Gulf on that side. On the south-east lies a fertile strip of level country, afterwards widening out into the malarious plain of Paestum; but, as a part of the Gulf separates this marsh from Salerno, the town is shielded from its dangerous effluvia by the intervention of salt water. I was informed by the intelligent chemist of the place that intermittent fever is rare in winter; but in summer the whole tract of flat country becomes malarious; and then fever prevails among the inhabitants. The climate is mild and sunny; and from the locality being situated on the sea-coast the air is stimulating. The town is only two hours by rail from Naples; and is one of the cleanest and least smelling of those that I visited in South Italy. The Hôtel d'Angleterre is excellent: and many delightful excursions by boat and carriage can be made into the surrounding country; always avoiding the popular but rather dangerous one to Paestum.

Proceeding along the lovely coast-road on the north side of the Gulf, we pass the picturesque villages of the Eastern Costiera, some day destined to be used as winter stations, and reach Amalfi, a town placed on the southern shore of the Sorrentine promontory, with mountains rising boldly in the background, and protecting it from northerly winds. It is built at the mouth of the narrow Valle de' Molini, ending in, but not penetrating, the St. Angelo range. Down this gorge a torrent, which turns a number of maccaroni and paper mills, descends to the sea, and impairs the otherwise complete shelter of the place, which has rocky masses rising north, east, and west, and is only open to the south, from which quarter it occasionally experiences stormy weather. The climate is probably warmer than that of Salerno; and excursions by land and water can be made into its beautiful neighbourhood, the scenes and brilliant colouring of which have been made so familiar to us by the labours of many great artists. The maccaroni here is excellent; and the Hôtel Capucini is comfortable, though placed in rather a noisy spot.

The northern side of the Sorrentine promontory is open to the full influence of the Tramontana; and therefore its numerous towns, Sorrento, Vico, Massa, &c., furnish agreeable and cool retreats in the summer; but their northern exposure, and their being shut out from the south by the St. Angelo range, render them unfit for winter resorts for pulmonary patients.

The accommodation of Sorrento is good, the food excellent, and the excursions in the neighbourhood numerous and charming; but the hotels and houses are built facing the north, and not so well suited for winter as for summer. Generally speaking, it is advisable for pulmonary invalids to return to England at the end of the spring, and reap the advantage of a cool summer and more generous living. Should a patient, however, find it inconvenient to do so, and prefer to linger near his southern refuge, he will find in the beautiful *Piano di Sorrento*, the vast garden of oranges and vines in which the town lies, a locality for a summer residence,* as enjoyable as it is healthy; combining the cool breezes of the Bay of Naples with its lovely views. The plain is three miles in length, and at an elevation of about 300 feet above the Mediterranean. It is besprinkled with villas, and from the varied colours of its vegetation forms a beautiful prospect from the adjoining hills.

We must not close this notice of the Italian winter stations without a parting tribute to the beautiful sea, which by its genial warmth contributes so powerfully to the mildness of their climate, and forms so important an element of their healing influence. The Mediterranean, on account of the deeply indented character of the Italian coast line, and the numerous headlands and creeks which its waters lave, is closely intermingled with its sunny shores; and from its high temperature throughout the winter, they derive as much warmth as they do beauty from its exquisite and ever-varying hues.

* For other summer stations see Appendix.

CHAPTER VII.

OTHER MEDITERRANEAN WINTER STATIONS: AJACCIO—MALAGA—TANGIERS—ALGIERS.

We have hitherto, with few exceptions, confined our attention to the winter stations of the French and Italian coast line, but it has been represented to us that a description of some of the other Mediterranean sanitaria, such as are to be found in Spain and Africa, and whose climate is connected by so many links of resemblance with that of the Riviera, would prove very acceptable to invalids, and render the book more complete. Our limits, however, do not admit of more than a few notices, in which we have endeavoured to sketch the most important features of each place in an impartial spirit.

Of late years, the mountainous island of Corsica has attracted much notice among climatologists, who have considered that its position in the Mediterranean, between 41° and 43° north latitude, and its rocky conformation, including ranges of considerable height and sheltered harbours, indicated it as a likely sanitarium. *Ajaccio*, on the west coast, has been recommended by Drs. Bennet and Ribton as a winter station, and has been resorted to by several invalids. The bay of that name has a southerly direction, and is protected both by the snow-covered central range of Corsica itself, and by its lateral spurs which run down on either side of the harbour. Its calm atmosphere and exemption from stormy blasts, especially in winter, have caused it to be regarded as a very safe haven by sailors, and the same qualities have attracted invalids to its shores. The town of Ajaccio is built on a small triangular promontory in the most sheltered part of the bay, and faces east towards the mountains, thus enjoying much protection, as proved by the vegetation; olives, oranges, and even lemon trees flourish, thickets of myrtle, arbutus, cistus, and Mediterranean heath, called *maquis*, abound in the neighbourhood, and magnificent chesnut groves are seen in the island. The scenery is admitted to be very beautiful, the combination of sea, mountain, and wood lending great charm to the landscape. The climate seems to be mild and remarkably free from cold winds, but the observations that have been hitherto taken do not extend over a sufficiently long period to warrant their being considered decisive. From Dr. Pietra Santa's report to the French government,* we find that, according to some observations of M. Nosadowski, carried on for five years (1854–8), the mean winter temperature was 53° Fahr. and the difference between winter and spring mean temperatures 5·57° Fahr. M. Dupeyrat kept a three years' record of the days on which rain fell, and found the annual average to be 48. From these scanty particulars we should not be justified in making a comparison with other winter stations, where meteorological records have been kept for a long period, but from these, and from the evidence given by vegetation, we may conclude that Ajaccio enjoys a mild winter climate. On the eastern coast of Corsica, and particularly near Corte and Aleria, intermittent fever is rife in summer and autumn.

The nature of the shore and the causes of the malaria are thus explained by Dr. Bennet:—'The eastern range, composed, as stated, of secondary calcareous rocks, is more easily disintegrated and washed away by the action of the elements. Owing to this cause the rivers which descend from its sides, and from the central regions of the island, through clefts which these calcareous mountains present, have deposited at their base alluvial plains of considerable

'* La Corse et la Station d'Ajaccio.'

extent. Through these rich alluvial plains several large streams meander to reach the sea. This they accomplish with difficulty, owing to the lowness of the shore, and the prevalence of the sirocco, or southeast wind, which constantly throws up large masses of sand at their mouths. Hence the formation, along the eastern shore, of large salt-water ponds, into which some of these rivers empty themselves. Under the burning glare of a Mediterranean sun, these terrestrial conditions—large ponds of brackish water, marshes, and rich alluvial plains, liable to periodical overflow—embody all the elements calculated to produce malaria of the most deadly character. By such malaria is this region rendered all but uninhabitable for half the year.' The eastern coast is the principal seat of this fever, but it exists in parts of the western, as near Calvi, and even Ajaccio is not quite free from it in early autumn, particularly when the wind blows from the mouths of the two rivers which empty themselves into the bay. Ajaccio contains several hotels, but from all accounts the accommodation, though tolerable, is very inferior to that of Nice or Cannes. Game and fish are said to be abundant, but mutton is scarce.

Malaga, the favourite wintering station of Spain, is situated on the Mediterranean coast, about 65 miles east-north-east of Gibraltar, and can be reached by rail in 48 hours from Paris. The town, containing more than 100,000 inhabitants, faces south-east, and is placed on the shores of a fine bay, four or five miles distant from the lofty mountains, which nearly encircle the rich and fertile plain in which the town lies. The semicircular barrier thus formed consists of two ranges. The nearer one is richly clothed with vines, and reaches an elevation of 3000 feet, sending off a spur to the shore, on the east side of the town, and thus affording some protection in that quarter; and again, on the west side, forming a protecting point, Torre Molinos, the western boundary of the Bay of Malaga. The loftier range, composed of the Sierras Nevada, Ronda, and Antequera, is bare and partially snow-clad, and separated by deep valleys from the vine-clad one, but together they form a double rampart of mountains, which, though not situated close to the town, affords it immense protection. The town is thus shielded from all northerly winds, except the *terral*, or north-west wind, a dry searching blast, much resembling the mistral in character, which reaches the town, through a deep fissure in the Antequera mountains called the *Bocca del Asno*, through which the river Guadalhorce flows to the sea. Malaga is open to all the southerly winds;

the south-east or sirocco, from its passing over eighty miles of sea, is moister than on the African coast; and the south-west is cold, and owing to its bringing storms from the Atlantic, is called the *vendebal*; the west wind seldom blows, but the east, or *levante*, is more prevalent, and being moist and chilly in winter, is much dreaded by the inhabitants. According to ten years' observations of Dr. Shortcliffe, the resident English physician, the mean winter temperature is 56° Fahr., the mean daily range small, and the difference between spring and winter mean temperatures does not exceed 6°. The rainfall is $16\frac{1}{2}$ inches, and the number of rainy days 40. The climate is warm and dry, its mean winter temperature being considerably higher and its rainfall considerably less than those of any winter station of Southern France. The transition from winter to spring is less marked than at Mentone, though more so than at Algiers. The vegetation is rich and varied, and, according to Dr. Bennet, shows evidence of a very mild southern winter.[*] Most authorities agree that the climate is one of the mildest in Europe, and well suited to pulmonary invalids, but, unfortunately, there are many hygienic

---

[*] The 'Gardener's Chronicle,' July 24, 1869.

drawbacks to Malaga as a winter station. The large population is closely packed, the streets are narrow, and the drainage, according to Dr. Madden, who has written an excellent account of the place, is most imperfect; the sewage is conducted in open channels on to the dry bed of the River Guadalmedina, where, under the sun's influence, it exhales unsavoury and noxious vapours. Again, the hotels are gloomy and uncomfortable, and the sheltered promenades are very limited, and as might be expected in so dry a climate, the dust is excessive. Dr. Madden says that the living includes a fair supply of fish, fowl, and game, but the meat is too tough for invalids, and the cooking being extremely bad, is not likely to render it more palatable. It will be seen that many of these drawbacks would disappear if villa residences were provided; and it is to be hoped, for the sake of invalids, that ere long this want will be supplied, for it were a pity that this spot, blessed with so fair a clime, whose natural advantages clearly fit it for a winter station, should be forbidden to invalids, on account of its disgraceful sanitary state.

From accounts given by several residents we gather that *Tangiers*, in Morocco, thirty-eight miles distant from Gibraltar, has strong claims to be recommended as a winter station for invalids. The town is built in the form of an amphitheatre, on the side of a hill, the summit of which is crowned by the Casbah, or Moorish castle. If vegetation be any test of climate, that of Tangiers and its neighbourhood would rank very high, for sun and soil have combined to render it one of the most productive regions in the world, and a great difference in fertility is observed in passing from Spain to this coast. Two crops of wheat are reaped in the year, the grain yielding an increase of from 25 to 35 fold, and oranges, lemons, figs, grapes, melons and dates are produced abundantly, and in great perfection. On the features of the climate Dr. Madden makes the following remarks :—'The position of the town exposes it completely to cold damp winds, which rush through the funnel-shaped straits from the Atlantic, while its aspect, being open to the east, must render it far more subject to this wind than Gibraltar is. The annual rainfall is about thirty inches, which, as in most parts of Africa, principally occurs at one season, during the months of October and November. The rains being succeeded by great heat, vegetation is consequently rapid and early; thus in January the fields are already covered with flowers, and in March the barley crops are reaped. Though the climate is hot, it is not parched or arid, as the province of El-Garb is protected by the interposition of the two ranges of the Greater and Lesser Atlas mountains, on the south and south-east, from the hot winds of the desert; its proximity to the Mediterranean and Atlantic on the north and west also modifying the temperature, which in this province seldom falls below 40° Fahr. in winter, or rises above 86° Fahr. in summer, at which season it should be a comparatively cool and agreeable residence.'

We have received from several English residents, who have had experience of other climates, a report much more favourable than the preceding quotation would represent, so far as regards the winter season. Protected as it is from the scorching blasts of the great African desert, and from many of the winds which sweep over the Mediterranean, Tangiers enjoys a softer and more tempered climate than any part of Algeria, resembling in some degree that of Madeira, but without the disadvantages of its relaxing influences and its remote isolation. It is easily accessible by the regular steam-packet service to Gibraltar, and it might be more confidently recommended if there were sufficient accommodation for visitors. This, however, is very limited, and is chiefly confined to the newly established Royal

Victoria Hotel, which, in addition to comfortable domestic arrangements, has the advantage of large gardens facing the Atlantic, forming a good promenade near the sea. Provisions, including the meat, are reported to be both excellent and moderate. The environs of Tangiers are beautiful, and embrace extensive views of the Spanish coast: but owing to the unsettled state of the country, it is not considered safe to wander far from the town without military escort, a circumstance which will considerably interfere with Tangiers becoming a place of general resort.

The French territory of Algeria lies between the 32nd and 37th parallels of north latitude; bounded on the east by the regency of Tripoli, and on the west by the empire of Morocco. Its northern limit is the Mediterranean, which washes its indented shores for upwards of 600 miles; while its southern borders extend into the Great Sahara desert, some of whose oases it includes. The mountain ranges of the Greater and Lesser Atlas traverse the province from west to east, and form two chains more or less parallel to each other for part of their course; but here and there giving off numerous spurs, which either subdivide the intervening country into valleys and plains of irregular shape and extent, or else run

INFLUENCE OF THE SAHARA.

north or south towards the sea or desert. These mountains attain the height of more than 7000 feet in the Warensis, the loftiest of the chain, and a great portion of them being snow-capped in winter, they form an important barrier whereby the land lying to the north of them is partially screened from the scorching influence of the sirocco. Dr. Henry Bennet, of Mentone, who has lately visited Algeria, with a view to observe its spring vegetation, considers that the climate of Algeria is greatly influenced by the neighbourhood of the Sahara, and in the following way:— The atmosphere which lies on this immense rainless tract, or desert, becoming heated both in winter and in summer, must rise into the higher atmospheric regions, and thus form a vacuum which the cooler and heavier air of the Mediterranean basin rushes down to fill. The latter is thus positively 'sucked in' over the summits of the mountain region of the northern shore of the Atlas range. Consequently, in Algeria, the regular winds must be and are either north-east or north-west winds, and south winds can and do only reign exceptionally. These northerly winds coming from the sea, or the ocean, are moist winds, and when they come in contact with the Atlas hills and mountains on the very shore, are at once in winter cooled, and deposit their moisture in copious and frequent rain over the entire Algerine or Atlas region, and right into the desert of the Sahara itself, for 250 miles or more from the sea.*

The favoured position of the country, the great varieties of climate, which the differences of elevation, and the neighbourhood of the sea on the one hand, and of the desert on the other, give rise to, have naturally pointed out Algeria to medical men as a region commanding great resources in climate; and the experience of Drs. Bertherand, Bodichon, and Pietra Santa, testify to its favourable effects, particularly in diseases of the respiratory organs. The presence of marshes in parts of the province, e.g. in the plain of the Metidja and those near Bona, are accounted for by Dr. Scoresby Jackson (in his excellent work,† to which I am indebted for much information about Algeria and some other localities,) as the effect of the rivers, which, though dry in the summer, increase enormously in volume during the rainy seasons of autumn and winter and overflow their banks, converting the neighbouring plains into noisome marshes.

Algiers, which is reached by steam in 48 hours from Marseilles, lies on the western shore of a beau-

---

* 'Gardener's Chronicle,' June 12. 1869.
† 'Medical Climatology,' page 96.

tiful crescent-shaped bay, the horns of which are formed by the Cape Matifou on the east, and the Pointe Pescade on the west. The town faces east and north-east, being built on the slopes of the Sahel range in the form of a triangle, the base of which is washed by the sea; and from the brilliant white of its houses and walls contrasting with the deep green margin of the surrounding slopes, it has been compared, not inaptly, to a diamond set in emeralds. The Sahel hills line the coast to some distance to the west of Algiers, rising to the height of 600 feet, and affording considerable shelter to the town from north-westerly winds. They are separated from the Atlas chains by the wide plain of the Metidja, which, commencing at the eastern shore of the bay, sweeps round the Sahel hills and rejoins the sea near Cherchell. This extensive plain, from 10 to 15 miles in breadth, like the Roman campagna, was formerly richly cultivated, and had a large population, but later on was deserted; and the parts near the coast became swampy and malarious. Since the French occupation, much has been done to reclaim this waste; and large tracts, including the Lake Halloula, have been drained, cultivated, and rendered more healthy. Still there are malarious portions which are to be carefully avoided, but in the suburb of Mustapha Supérieur, and on the slopes of the Bouzarèah, the highest of the Sahel range, are to be found numerous villas and other residences fit for invalids. The year is divided into two seasons, the hot and the rainy, the former extending from April to November, and the latter from November to April, the greatest rainfall occurring in the months of November, December, and January. Dr. Pietra Santa, who was appointed by the French Government to report on the climate of Algeria, states, as the result of 22 years' observations, from 1838 to 1859 inclusive, the winter mean temperature to be 56° Fahr.*; the amount of rainfall 32·18 inches; and the number of rainy days 87. For the last five years the rainfall has been much less, and the dry seasons have been the cause of famine and plagues of locusts. The mean temperature of the whole rainy season is 62°, and the differences between the winter and spring mean temperatures are small. The most prevalent winds seem to be westerly, and are the west, the north-west, and the south-west; all of which contain moisture, and temper the extreme heat. The sirocco is here a terrible wind, injurious to man, and disas-

* Dr. Scoresby Jackson gives, as the result of 13 years' observations, a mean temperature of 62°, which is much higher than the above.

trous to vegetation; but I am informed by Dr. Stewart Gentle, the resident English physician at Algiers, that it prevails from June to the beginning of October, and that it only blows mildly during the temperate months, and therefore does not much affect invalids. The rainfall of Algeria, curious to say, increases on proceeding eastwards; for, of the three provinces into which the country is divided, Oran, the most westerly, has the least rainfall; Constantine, the easternmost, has the greatest; while in Algiers, the central one, the rainfall is double that of Oran, and about half that of Constantine. This may possibly be accounted for by the distribution of forests, which are most extensive in the province of Constantine, but, from various causes, have nearly disappeared from Oran, and are much diminished in Algiers. The climate is warm and stimulating, but differs from that of the north coast of the Mediterranean in many respects; the mean temperature being much higher, and the change from winter to spring being less marked. These features are not to be accounted for by the shelter of mountain ranges, for, though the Sahel hills give some protection on the west and north-west, the Atlas chain lies too distant to afford much shelter, but it is in the more southern latitude, tempered by the equalizing influence of the Mediterranean, that an explanation may be found of this mild climate. The fact of the rainy season occurring in the winter, instead of in the autumn months, as in most of the North Mediterranean stations, is certainly a drawback for invalids. The number of rainy days, too, is great; but according to Dr. Scoresby Jackson, the rain does not last long; and what is called meteorologically 'a rainy day' at Algiers may in fact mean nothing more than a very heavy shower, of half an hour's or an hour's duration, not materially interfering with out-door exercise. Such brief showers have the beneficial effect of laying the dust, which is much complained of in Algiers in dry weather.

All observers agree that the climate is eminently suited for chronic bronchitis and for the non-inflammatory form of consumption, but on the inflammatory form it seems to have similar effects to that of Mentone. Of the hotels of Algiers, the Hôtel d'Orient is near the sea, and faces south-east, across the bay, as also does the Hôtel de l'Europe; while the Hôtel de la Régence looks towards the south, and is therefore warmer, but its drainage is complained of. The excursions into the Sahel and Atlas ranges are varied and charming, and many are of a nature to be undertaken by invalids. At some of the hill stations, as Blidah, Milianah, and Medeah, tolerable hotel accommodation is to be had; and the climate, as evidenced by the vegetation, is very mild. In excursions in the Metidja plain care must be taken to avoid the marshy parts, for intermittent fever is far from rare, and cases occur of a severe type. For the same reason Mustapha Inférieur is to be avoided as a residence, as it lies low and is swampy.

APPENDIX.

'Whereas according to researches of Ross, Belcher,' etc., p. 20.

Mr. Gwyn Jeffreys, F.R.S., who has lately been engaged in important deep sea dredgings off the west coast of Ireland, has carried out a very interesting series of observations on the temperature at great depths. He informs me that the results of previous thermometrical investigations were more or less erroneous, on account of the bulb of the thermometer not being protected from the pressure which the mass of superincumbent water exercises on it. This defect has been remedied by Dr. Miller, F.R.S., who has enclosed Six's self-registering instrument in a second glass tube, the interval being filled with spirit; and thus variations in external pressure are prevented from affecting the bulb of the thermometer, within, whilst changes of temperature in the surrounding medium are speedily transmitted through the interposed alcohol.

It appears from reliable observations, made with this instrument by Mr. Jeffreys, off the Rockall Bank to the north-west of Ireland, and by Professor Wyville Thompson, in the Bay of Biscay, that the temperature at the depth of 1476 fathoms is 36·9° Fahr., and at that of 2435 fathoms (nearly three miles) 36° Fahr.; and that at these great depths there is abundant evidence of animal life, but that the Fauna, as might be expected from the temperature, is chiefly of the arctic description. Further researches in the Mediterranean and Atlantic, with instruments made on Dr. Miller's plan, are required before we can arrive at just conclusions as to the relative temperatures of these seas.

Summer Stations.

However beneficial warm sheltered localities may be to pulmonary invalids in winter, they become oppressive and enervating in summer; and it is therefore usual to recommend a removal to a cooler and more bracing climate; where the appetite, often impaired by a long spell of warm weather, may be rendered keener, the nervous system refreshed by sounder sleep than is usually procurable on the Mediterranean coast, and the muscular power invigorated by better food, and by increased exercise.

The majority of English consumptive patients, after an exile of so many months, prefer to return to their native land, whose moist climate seems an agreeable change from the dry air of the south, and whose wholesome and nutritious food forms a great contrast to that of most foreign countries.

Many invalids however, who intend to pass a second winter on the Mediterranean, wish to save the expense of a journey home, and would prefer to pass the summer in some cool mountain retreat, at no great distance from their wintering place.

The Maritime Alps offer advantages in this respect, and the following mountain retreats, of which the descriptions have been kindly forwarded to me by Drs. C. H. Battersby and Daubeny, may prove available.

'*St. D'Almas de Tende* lies 50 miles distant from Nice, on the high road to Turin, about 2 miles on the Italian side of the frontier, in a wide valley formed by the junction of three mountain streams, which uniting here form the river Roya. The four valleys thus formed afford a free circulation of air, so often absent in an unbroken mountain valley, whilst the surrounding hills break the force of the winds. St. Dalmas is particularly well protected, on the north by the range of the Col de Tende, and owing to its altitude, nearly 3000 feet, and the neighbourhood of mountains, 9000 feet high, covered with snow for the greater part of the year, the nights at St. Dalmas are always cool; and even in the hottest seasons there are few days on which it is not pleasant to go out in the middle of the day. The mean temperature of the summer months on an average of 6 years is as follows: June 65° Fahr., July 69° Fahr., August 68° Fahr., and September 63° Fahr.

'The minimum in June and September is about 50° Fahr., and the maximum in July and August about 84° Fahr.: these limits are however seldom reached. There are frequent showers during the summer; but notwithstanding this, and the abundance of running water in every direction, the air seldom feels damp, and there is no evening dew. The soil being limestone gravel, the roads dry so quickly that one can walk with pleasure as soon as the rain is over. The climate is very healthy, as is evidenced by the state of the peasantry, who are noted for being the most robust in the department, and for some years, since the irrigation has been conducted on a better system, there have been no cases of intermittent fever, of which there were complaints at one time. Delicate children, and persons suffering from various diseases, appear to derive much benefit from a sojourn at St. Dalmas, regaining appetite and strength very

K

rapidly. The Hotel and Pension, formerly a Benedictine monastery, is well managed by the proprietor of the Hôtel des Princes at Nice, and the terms are moderate. It is very pleasantly situated, partly surrounded by chesnut groves, with a large and well wooded garden, sloping down to the river, affording level and shady walks for those who are unable to undertake the longer excursions, which are almost exhaustless, and interesting both on account of the lovely scenery, and the varied nature of the botany and geology.

'C. H. BATTERSBY.'

Another summer station is *Certosa di Pesio*, in the Maritime Alps, about two and a half days' journey from San Remo, of which Dr. Daubeny has kindly furnished me with the following notice:—

'Certosa di Pesio is an old Carthusian convent on the river Pesio, built about the year 1450, and for the last three years used as a pension by the Signori Boggi, who are the proprietors of the Hôtels de Londres at San Remo and at Mentone. The house is well furnished, and will accommodate about 100 visitors, exclusive of servants. As a summer residence it is admirably situated in a very beautifully wooded park; at an altitude of about 5000 feet, a grand torrent stream, which supplies excellent trout fishing, running through the grounds. During the hot summer months, there are always plenty of shady walks, while an Hydropathic Establishment there contains a great variety of baths, which can be used either for health or luxury.

The journey from Oneglia, the point where the road diverges from the Riviera route, takes two days in a due northerly direction. The hotels on the line, if properly chosen, are tolerable. All who contemplate this journey should apply to Signor Boggi at San Remo, or at Mentone, for information. Certosa can likewise be approached from Turin by rail to Cuneo (two hours) and thence by half a day's drive to the Convent. The time for living at this place is from the first week in June, until the beginning or end of September, according to the season. The Hotel is well conducted, and there is a great variety of outdoor and indoor amusements for visitors, including fishing and shooting.

'I visited Certosa in June and found the climate deliciously cool and invigorating after that of the Mediterranean.

'HENRY DAUBENY.'

The valley of the *Upper Engadine* in the Swiss Canton of the Grisons is now much resorted to in summer, partly on account of its high altitude and consequent coolness, and partly on account of the virtues of the mineral waters of St. Moritz, one of its villages, situated 6000 feet above the level of the sea. The valley, one of the highest inhabited spots in Europe, is wide and traversed by the river Inn, here forming a series of picturesque lakes, over which tower the mountains of the Bernina range, their sides clothed with firs, and their summits crowned with eternal snow and ice.

The principal villages of this region are St. Moritz, Samaden, and Pontresina, and these all contain tolerable hotel accommodation for the summer season, which extends over barely three months. The atmosphere is clear, and the air is dry and bracing, but although the sun's rays are powerful and scorching during the day, the nights are very cold, and sometimes even frosty. These extremes of temperature amount to as much as 50° Fahr. in the 24 hours, and are not without their evil results; sometimes giving rise to feverish attacks and inflammations. In fact the upper Engadine is by no means a safe summer residence for pulmonary invalids. Those who thrive in a keen bracing air may go there with advantage; but they should be prepared with suitable clothing for the variations in temperature which they may encounter; and they should moreover make sure that they can get comfortable accommodation in the best hotels, which is by no means an easy matter, now that the Engadine has become such a fashionable summer resort. The Bernina Hotel at Samaden is the best in the Engadine, and is the only one open through the winter. The Kurhaus and Kulm at St. Moritz, and Krone at Pontresina, are much frequented in the summer season.

For the plurality of invalids who require a bracing and moderately cool situation during the summer months, several places, about a thousand feet lower, will be found more genial than the upper Engadine. Tarasp-schuls, a large bathing establishment in the upper part of the lower Engadine, with excellent accommodation for three hundred guests, at a height of 4000 feet above the sea; Davos am Platz, 5000 feet, with a large Kurhaus, where inmates are comfortably lodged on moderate terms; and the Nuovi Bagni of Bormio, on the Italian side of the Stelvio pass, at a height of 4569 feet,—are all high, dry and healthy summer residences, in which the extremes of the upper Engadine are avoided, whilst the invigorating air and ever varying beauties of Alpine heights may be enjoyed in perfection. The last mentioned place, the new baths of

Bormio, especially deserves commendation. Situated at the head of the upper Valtelline, commanding the grand vista of that beautiful gorge, with two other valleys to the right and to the left, with their snow-topped peaks, and backed and sheltered from the north and east by the huge masses of Monte Cristallo and Piz Umbrail, the great mountains of the Stelvio, the highest pass in the Alps—it presents a rare combination of airiness and shelter, of dryness and luxuriant vegetation, of salubrity and beauty, not met with in any other place. The establishment receives two hundred invalids, and is to be further enlarged. The thermal springs supply warm and mud baths of various kinds, which are much in request during the season, which extends from the middle of June to the middle of September. Bormio may be reached either over the Stelvio pass, which is now in good repair; or from the Italian side, by the road up the Valtelline, from Como, or from the Bernina.

Another charming summer retreat may be found at Le Prese, in the valley of Poschiavo, on the descent from the Bernina pass. This is another bathing establishment, the water being sulphurous and alkaline: but the hotel accommodation is also excellent, being managed by M. Mella, formerly of the Grande-Bretagne at Bellagio, and who made that delightful hotel so attractive to travellers. The height of Le Prese is only 3000 feet, and it cannot be equally bracing with those already mentioned; but it forms a good intermediate between them and the sites on the lakes of Como and Maggiore.

In Switzerland, also, are several comfortable Hotels at great heights, which are available during the summer for those patients likely to benefit by the Alpine air; they are generally more accessible and less liable to extremes than the Engadine. Such is Leukerbad, above the Rhone Valley, at a height of 4642 feet; Comballaz, above Aigle, in the Lower Rhone Valley, 4416; Ormond Dessus, under the Diablerets, 3832; and the Rigi Kaltbad, 4728.

The Engadine, I may remark, has lately been recommended by Dr. Hermann Weber and others as a winter residence for consumptive invalids, on account of the exemption from consumption which its inhabitants are said to enjoy. For this Dr. Weber assigns as a cause the high altitude, and the state of atmosphere consequent thereon, and maintains that cold air, when dry, is not injurious to consumptive invalids, but rather beneficial. He gives an account of 31 cases which have been subjected to this treatment, and states that in the greater number of instances the disease was lessened and that blood-spitting was arrested. The cases have not been narrated fully enough to enable one to judge of the results of the climate, nor is their number (31 in all) large enough to justify conclusions being based on them. With regard to the theory that dry cold is not injurious to invalids, it is possible that by bracing the constitutional powers it may exercise a good effect on individuals predisposed to Phthisis: but where the disease has already developed itself, severe cold, even when dry, is apt to induce catarrhal and inflammatory complaints, and would thus render the patients considerably worse. A perfectly *still* cold atmosphere is less harmful than one in which winds prevail; but, according to accounts, the Engadine is by no means free from wind, particularly in the winter months.

APPENDIX II.

FURTHER OBSERVATIONS ON SUMMER QUARTERS FOR INVALIDS, AND ON ALPINE SANATORIA IN 1870.

When it is considered what large numbers of British travellers resort every summer to Switzerland and other mountainous countries for the improvement of health as well as for enjoyment of scenery, it may not be out of place to add a few remarks on the summer quarters most suitable for invalids, and the considerations which ought to guide in their choice.

That mountain districts, and high and dry situations generally, are more healthy than those which are low and damp, is too well established to require further discussion. In Switzerland and North Italy no traveller can fail to observe the striking contrast presented by the inhabitants of the low deep valleys, as compared with those of the mountains and elevated plains. Even the borders of the lakes, with their verdant banks, overhung by mountains, and presenting the most enchanting combinations of scenery that can delight the eye, are not favourable to the development or preservation of vigorous health. The inhabitants have generally pallid or sallow complexions, and goîtrous throats; and the visitor soon finds his enjoyment to assume the passive form of sailing on the lakes, and driving, or lounging, instead of the enterprising activity and elasticity, recreating and invigorating, of the mountain climber.

Those travellers whose object is not only beauty of scene but improvement of health, will do well to spend but a small portion of their time, and only such as stress of weather or fatigue may require, in the lower levels, and at once betake themselves to higher altitudes, where they may experience the salutary effects of the dry pure atmosphere peculiar to those regions. These qualities of the atmosphere vary much according to the height and position of the place, affording us a great choice of situations suitable for different classes of invalids; and on the discrimination with which the locality is selected may much depend the wellbeing of the patient.

Alpine resorts may be arranged, according to height, in three classes:—

I. Towns and villages at a comparatively low level, situated on the borders of large lakes, as Zurich, Lucerne, Brienz, Thun, Neuchâtel, Geneva, Vevey, Montreux, and Evian; or on plains not much raised above the lakes, as Interlaken, Glarus, Martigny, Berne, &c.

The majority of these are more or less damp, and in summer often oppressively hot, and in several the drainage is very defective. Geneva has the advantage of the purifying and ventilating influences of the rapid Rhone sweeping through it, and to a less degree Lucerne and Thun are similarly favoured. On the other hand, Aigle, Bex, Martigny, Sion, Brieg, and other places in the deep Rhone valley, have so much flat and marshy ground near them, as to be positively malarious, and the sooner travellers can get out of them the better.

II. Moderately elevated situations, at heights ranging from 2000 to 4000 feet, where mountain air can be enjoyed, though not so bracing in character as that of the highest stations; but being less liable to extremes of temperature, these places are more suitable for first trials and for delicate persons, especially those unaccustomed to mountain climates. Many of these, too, present advantages in superior accommodation and ready means of access, which make them more available for the greater number of invalids.

Over the lake of Geneva, the most attractive mountain abode is Glyon, above Montreux, at a height of 3000 feet above the sea. Comfortable quarters may be found at the Hôtel du Rigi Vaudois and other hotels; for the magnificent scenery of the east end of the lake makes Glyon a favourite resort. At the opposite end of the lake, over Nyon, at a height of between 3000 and 4000 feet, up the Jura range, is St. Cergues, which in clear weather commands one of the grandest distant views of the Mont Blanc range to be seen in Switzerland.

The following are other instances of this class :—

Champèry, 3412 feet above the level of the sea, situated close under the Dent du Midi and the Tour Saillière, about nine miles from the Monthey station of the Rhone Valley railway, has good hotel accommodation, and abounds in interesting scenery for excursions.

In the beautiful and easily accessible Diablerets district are,—Sepey, 3300 feet above the sea, 7½ miles from the Aigle railway station, with fair hotels—some open during winter. Three miles farther, Comballaz, 4416 feet above the sea, with good and moderate boarding-houses; and at the head of the valley, another good hotel, at Ormond Dessus, 3832 feet above the sea. This Val des Ormonds, rich and wooded at its lower end, rises to grandeur at its upper extreme, in the Creux des Champs, under the snowy summits of the Diablerets, and contains as great a variety of scenes, gentle and savage, as the eye can desire.

In a visit during June of this year (1870), I found Sepey very hot, notwithstanding its height; Vers l'Église, in Ormond Dessus, and Comballaz, cooler and more suited for summer quarters, and fairly sheltered from the coldest winds; Plan des Isles, also cool, but rather marshy from its numerous rivulets. Above, on drier ground, stands the Hôtel des Diablerets: in this, and several other hotels in this district, visitors are boarded at from three and a half to five francs a day; food good, but furniture scanty.

An der Lenk, 3527 feet above the sea, six hours from Thun, lies at the head of the Simmenthal, on the northwest side of the snow-capped Wildstrubel—the valley being closed with glaciers—and the visitor will find good accommodation and no lack of interesting excursions in the neighbourhood. I have also heard a good account of the Krone at Zweisimmen, at a height of about 4000 feet in the Simmenthal.

Engelberg, 3300 feet high, is beautifully placed in a valley on the north side of the Titlis, with a large new hotel, much resorted to for health as well as for the scenery. It can be reached in a few hours' drive from Stanzstad, on the Lake of Lucerne. On the same lake, and accessible by the steam-boat station of Treib, and above Seelisberg, is Sonnenberg, with two good boarding-houses, at a height of 2759 feet, overlooking the Bay of Uri, with many points of interest—commanding magnificent views of the whole lake of the four cantons.

Grindelwald (3737), with good hotel comforts, and Lauterbrun (2595), near Interlaken, may be classed among the healthy summer quarters of moderate height and easy access; owing their coolness rather to the proximity of glaciers and the shade of their over-towering mountains, than to their elevation. The mountain inns above, Mürren (5347), Little Scheideck (6696), and Baths of Rosenlaui (4397) belong to the next class; but, however welcome as halting-places in scenes of surpassing grandeur, they are not to be numbered among places suitable for invalids.

III. High level resorts, ranging from 4000 feet upwards. In these the full bracing influence of mountain air is to be obtained; and where the powers of circulation are good, and the individual can endure the considerable changes of temperature which the heat of the sun's rays by day, and the chill in their absence by night, occasion, he may profit greatly by a stay at these heights, and find himself invigorated and able to undertake increased mental and bodily exertion. This is especially the case with those persons who have long suffered from the oppressive and relaxing air of lower valleys and plains. Pulmonary and other invalids may also improve by a sojourn at these altitudes, provided that in selecting the place of residence some considerations besides those of height and purity of air be attended to. Shelter from the coldest winds, the absence of swamp or stagnant water, a sunny aspect, a comfortable hotel and good food—should be secured to the invalid; otherwise he may suffer, instead of profit, by his high abode.

It is at such considerable elevations that it has been declared by several trustworthy observers that pulmonary consumption and kindred scrofulous diseases rarely occur; and that persons suffering from such maladies coming from lower regions into these altitudes, experience remarkable benefit, not merely in the mitigation of symptoms and improvement of present health, but in some instances amounting to the establishment of a curative process, by the removal of tubercles and inflammatory products, and a more or less complete restoration to health. The earliest notice of this curative power of high altitudes appears to be that of the late Dr. Archibald Smith, in a paper published in the 'Edinburgh Medical and Surgical Journal' in 1840.

In Lima and other towns in Peru, which lie low, pulmonary consumption is very prevalent, and it has been long the practice, popular as well as professional, to send invalids up into the Andes mountains, to an altitude of from 8000 to 10,000 feet, and with the most strikingly beneficial results. Dr. Smith quotes from Dr. Fuentes, of Lima, a statement that of patients in various stages of consumption sent from Lima to Jauja (an elevation of 10,000 feet) nearly 80 per cent. are cured. Dr. Williams has the notes of the case of a Peruvian gentleman who had become decidedly phthisical in this country, and returning to his native town, Lima, he had repeated attacks of dangerous hæmorrhage; in a state of great weakness he was carried up the mountains, and completely recovered; he remained free from pulmonary symptoms, and died of fever some years after. The observations of Guilbert on the mountains of Peru and

Bolivia, and those of Jourdanet, on the high plateau of Mexico, confirm the inferences of Dr. Archibald Smith on the curative power of great altitudes on phthisis.

In Europe Dr. Lombard long ago drew attention to the fact that phthisis is of comparatively rare occurrence in the higher habitable districts in Switzerland; and the positively curative power of high mountain air has recently been strongly advocated by Drs. Brehmer, Küchenmeister, and Spengler in Germany and Switzerland, and by Dr. Hermann Weber in this country. A brief allusion has already been made (p. 134) to Dr. Weber's paper on this subject, which has, since the printing of that portion of this little work, been published in the fifty-second volume of the 'Medico-Chirurgical Transactions,' and gives valuable information and suggestions 'On the Treatment of Phthisis by prolonged residence in elevated regions.'

Dr. Weber notices that the observed immunity of elevated regions with regard to phthisis varies greatly in different latitudes, requiring the greatest altitude—up to 9000 or 10,000 feet, within the tropics; not more than 3000 or 4000 feet in the South of Europe; whilst in the northern parts of Germany phthisis is said to be rare in places between 1500 and 2000 feet above the sea. Küchenmeister suggests an approximative calculation that for the latitude of Germany and Switzerland, the exemption falls about 375 feet for every degree of latitude from south to north. Dr. Weber questions the reality of any exact arithmetical relation between latitude and the altitude of the prevalence of consumption, and remarks that incongruities may be discovered even in the same mountain range. Thus he learns from Dr. Unger at Davos, that Dr. Boner of Klosters not rarely met with phthisis at Splügen (over 4700 feet), while it scarcely ever occurred in the upper part of Klosters (rather below 4000 feet). Dr. Weber very judiciously suggests that other conditions besides elevation are probably required to secure the inhabitants of a mountain district from liability to consumption; and amongst others he particularly specifies dryness of soil, referring to the observations of Bowdich and Buchanan, to which we have already adverted (p. 63).

Dr. Weber gives the history of seventeen cases of phthisis in various stages, which have more or less signally benefited by a residence at great elevations. The most satisfactory of these are those who had resorted to the high station of Jauja in the Andes, the cures in some of these instances remaining complete for several years. Other cases are those of German and Swiss artisans who have become phthisical on exchanging their country abodes for the close habitations and impure air of London, and recovered health on returning to their native homes in the Black Forest or in the Swiss mountains, but relapsed and fell victims to the disease on returning to London. These can hardly be taken in evidence of the peculiar efficacy of great elevations; for the heights resorted to were quite moderate, and the resulting temporary improvement not greater than that commonly experienced by workpeople who, after sickening in the unhealthy atmosphere and occupations of town, are restored by the pure air of their native homes in the country.

Another position assumed by Dr. Weber, which is open to question, being by no means established by the amount of evidence yet adduced in its support, is that phthisical patients benefit by residence at great elevations in winter even more than in summer, and that cold, however extreme, has not that injurious influence on such cases which it is commonly supposed to have. On this point it will require a larger amount of evidence, in form of numerous well-observed facts, to reverse the common opinion, that whatever may be the power of cold to prevent phthisis, yet when the disease is actually present, with its usual inflammatory complications, cold is a hurtful agent from which the great plurality of sufferers shrink with instinctive dread. There may be exceptions, but this seems to be the rule. In slight and incipient cases the case may be different; and all that has been said for the salubrity of mountain climate is in favour of its trial being first made in the summer, and with due regard to other influences besides that of altitude.

As a further illustration of the subject of mountain resorts for invalids, I have my father's permission to introduce a long extract from his 'Notes on Alpine Summer Quarters.'* They refer chiefly to the Engadine and Bormio.

'Our next halting-place was Thusis, at the foot of that wonderful triumph of engineering art, the Via Mala. Thusis is healthily placed—not so much from its absolute height (2448 feet), as from its being high and dry above the rich Domleschgerthal, and overtopped only on one side by the pine-clad rocks of the Via Mala. The small hotel there is also very comfortable. Travellers going to the Engadine, who wish to see the Via Mala, may well halt here, and afterwards join the Julier or Albula routes by the new road just opened through the Schyn-pass to Tiefenkasten, which is by no means a good place to stop at. At Tiefenkasten, we took the diligence over the Julier, to make a second trial of the Engadine. All that I had heard of the crowded state of St. Moritz and Pontresina determined me to try Samaden, where I knew the Bernina Hotel was sure to provide well in the way of food. Even here, although they have upwards of one hundred beds, and I had written a week before, we were obliged to be content with bedrooms in the village for two days. However, nothing could be more courteous than the conduct of the landlord, M. Franconi; and we were most comfortably lodged and fed afterwards, at prices quite moderate, considering the demand of the season. And let me say here, once for all, that the complaints which were continually reaching our ears of the short commons at some of the establishments at St. Moritz by no means applied to the Bernina at Samaden. In fact, we were assured that a good

* 'Notes on Alpine Summer Quarters for Invalids in 1869.' By Charles J. B. Williams, M.D., F.R.S. 'British Medical Journal,' November 1869.

many visitors used to come from St. Moritz to the Bernina Hotel for the sole purpose of getting a good dinner.

'The weather, which had been wet the whole preceding day, and was showery on our passage over the Julier, became fine on the descent; and we entered the Engadine in bright sunshine. Still there was a scattering of fresh snow far below the usual snow-line; and as we drove through Silvaplana, Campfer, and St. Moritz, we observed ladies wearing warm cloaks and even furs—a very sensible proceeding, but telling its tale of the climate in the middle of August. The weather continued generally fine during our week's stay at Samaden; but there were several falls of rain towards evening, succeeded by the appearance of fresh snow on the mountains on the following morning; and on four out of the seven nights, the grass of the valley of Samaden and of Upper Pontresina was white and crisp with frost in the early morning. Of course, all this vanished with the first approach of the sun's rays: in fact, so hot were they, that veils, white handkerchiefs, and umbrellas were brought into full requisition to avoid their scorching effect on the skin; and, even with these protections, few of the more zealous excursionists entirely escaped the branding and skinning of the face and neck, so familiar to holiday mountaineers. To many, these face burnings were trifles; and by giving a ruddy glow to a hitherto pallid cheek, encouraged the impression that they were signs of new health and vigour. No doubt, to many, the bracing air of the high Engadine is reviving and invigorating, and especially to those who have long suffered from the oppressive and relaxing air of the lower valleys and plains. These feel new life and energy in the cool bracing air, and their appetite and strength rapidly improve under its influence. With others, again, the effect is quite different: they are chilled without being braced, and scorched by the sun without being permanently warmed. They feel a certain degree of excitement in the air, but it causes fever, instead of strength; and its injurious operation is manifest in restless nights, and in failing, instead of improving, appetite for food. Our party, unfortunately, belonged to this latter class; and, consequently, the week which we spent in the Engadine was by no means a pleasant one; and we were not disposed to try a longer experiment of acclimatisation. We met with several others who made the same complaints. No doubt, the excursions in this neighbourhood are full of the beauty and interest attaching to the highest Alpine scenery. For example, Val Roseg is extremely picturesque and diversified in its pine-grown heights, forming an ever-varying framework to the dazzling glaciers and snow-peaks at its upper end. Considering that the woods consisted chiefly of two species only, the larch and the Alpine cedar, with only here and there a Scotch pine, there was a surprising absence of monotony; the grandeur of the rocky masses, and the brilliancy of the flowers and lichens, making up for the want of variety in the foliage. As usual in sunny days, the ascent was very hot, except where the way was shaded by the welcome trees; but the descent was from west to east, in the face of a cold wind, which became so cutting towards sunset, on our descent to the valley of Samaden, that we returned miserably chilled, in spite of all our wraps and the jolting of the rough berg-waggon, the only vehicle used for the by-roads.

'From these illustrations of the summer climate of the upper Engadine, it is pretty clear that it ought not to be recommended indiscriminately to all classes of invalids; and those that are likely to benefit by its exciting and bracing influences ought to be cautioned and prepared against its extremes, which are especially trying on first arrival. It appears to be the common experience of those who most benefit by the climate, that the powers of circulation and respiration so improve as to be able to bear the changes of cold and heat, which were found very trying at first; and this kind of acclimatisation may be extended to the colder season, so that those who have gone in the summer gain the power of so well enduring increasing cold, as to be able to pass through the almost Arctic winter of these heights without suffering or inconvenience.

'When it is authentically stated that several consumptive patients have passed the winter in the upper Engadine, not merely without suffering, but with considerable benefit to their general health, and with marked diminution of the symptoms and signs of disease in the lungs, the matter becomes one of fact, not of reasoning or opinion. But it is of the utmost importance to test the accuracy of the fact by adequate investigation. I have had no opportunity of personally examining any of the patients who are said to have benefited by passing the winter in the Engadine. Dr. Berry, the intelligent resident physician at St. Moritz, informed me that he had the history of ten or twelve such cases, several of which had been formerly under my care. My recollection of most of these is imperfect, but the following I have been able to identify.

'Mr. S., aged 20, consulted me on August 2nd, 1865. At the age of ten, he had glandular swellings in the neck, which gathered and remained open for twelve months. Last autumn, he was much reduced by a local inflammation, and leeching for it, and remained weakly through the winter, with a slight cough. In May, he coughed up three ounces of blood, and, a few days afterwards, half a pint. He was treated with lead first, and subsequently with iron. The bowels were often costive. I found some dulness and rough breath-sound in the upper right back. There was slight bronchophony above the right scapula. I prescribed for him cod-liver oil in a mixture containing phosphoric acid, hypo-phosphite of lime, and strychnia; an occasional aloetic pill; and tincture of iodine externally. To winter at Torquay.

'I heard nothing more of him till August 1869, when his father called on me at Samaden, and told me that his son had continued my treatment with considerable benefit, and passed the winter at Torquay. He had little cough, but there were occasional recurrences of hæmoptysis, and he several times expectorated calcareous matter.

'In June 1866, he went to Silvaplana, in the upper Engadine; and from that time there was no recurrence of hæmorrhage. The pulse gradually was reduced from 108 to 70, and the respirations from 28 to 18, with a corresponding improvement of strength and activity. The three subsequent winters he has passed at St. Moritz, going out daily, and enjoying skating, sledging, and other winter exercises, with impunity.

'In this case, there had never been much cough or other symptoms of bronchial or pulmonary irritation. The disease was in a quiescent state when he first went to the Engadine; and, going there in the summer season, there was time for acclimatisation before the severe weather set in. Such, in truth, seem to me to be the chief conditions most favourable to safety and success in trials of the *Alpine cure* of consumption. To send patients in advanced, or even recent, *active* disease, with its attendant local inflammations and congestions, and with the general weakness of circulation and no power of resistance against cold, to such a climate as that of the Engadine in the beginning of winter, does seem most rash and irrational. But, in cases where the pulmonary deposits are moderate in amount, and attended with little vascular irritation, and especially when the system is relaxed, and there is a consciousness of refreshment and invigoration in a dry cool air—under such circumstances a mountain residence offers the best promise.

'Such a condition, favourable for the *mountain cure*, is not unfrequently presented by pulmonary invalids who have passed the winter in a warm situation, such as Mentone, Cimiez, Hyères, or Pau. By these means, and by appropriate treatment, the disease has been brought into a quiet state; but the patient may have become more or less relaxed and weakened by the increasing heat of May and June. Then, with proper precautions in the way of clothing and choice of weather, the ascent to the higher region may be made with great probable benefit; and, should that benefit continue during the summer months, with improved strength and breath, and power to endure the changes of temperature occurring even at that season, there will be encouragement to prolong the stay through the winter, and thus prove the further efficacy of the mountain abode.

'And now a few words as to the selection of this mountain residence; for other points besides altitude require consideration. What absolute height—whether at 6000 or 5000 feet above the level of the sea—may be best, is yet to be determined by experience; but there can be little doubt that *shelter from the coldest winds, an aspect favourable to receive all the sunshine obtainable in the winter, a dry unswampy soil, and a comfortable house,* with well-regulated stoves to warm, and a well-supplied larder to feed, the winter inmates—*will be essential to their well-doing.*

'Of the villages in the upper Engadine—St. Moritz, Samaden, Campfer, Silvaplana, and Pontresina—all have hotels and boarding-houses; but the latter three of these are open only during the short summer season of three months. Pontresina, from its exposed situation, open to north and east, and its proximity to the snow-mountains and glaciers of the Bernina range, is wholly unsuited to invalids even in summer. Silvaplana and Campfer are sheltered from the north; but, being higher up the valley than St. Moritz, and between the lakes of the Inn, they are draughty, and liable to evening chills from the sheets of cold water. The Kurhaus of St. Moritz is in even a worse position; for it stands on a flat ground, very little above the lake. In fact, when I first visited it, seven years ago, this ground was little better than a swamp. Part of it has since been raised, and converted into a garden and croquet-ground; but the whole establishment is so far from the mountains which protect the village of St. Moritz to the north, is so much under the shade of the huge pine-clad Piz Rosatsch to the south, and so open over the lake to the east, and up the valley to the west, that it may be said to be open to every wind that blows; and, even in the summer, it must be both damp and draughty. The village of St. Moritz is much more favourably placed, being 400 feet above the lake, and 6085 above the sea, and considerably sheltered to the north by the mountains of the Julier pass, and to the east by a wooded hillock, which bars the valley, and forms the east bank of the lake. The hotels and boarding-houses here have more or less the advantage of this shelter, and of a declivity fully facing the south. The Rev. Mr. Strettell has built a comfortable house above any of the others on this slope, and he believes it to be the highest gentleman's residence in Europe. He told me that so warm are the sun's rays, even in winter, in this high yet sheltered spot, that the inmates have sometimes been able to sit with open windows, and even in the open air. The rooms are well warmed by means of the ordinary German stove; but this causes an excessive dryness and close feeling in the air, which is neither healthy nor agreeable. It is said that ventilation may be safely supplied by occasionally opening a window; and the extreme dryness may be prevented by placing on the stove shallow vessels of water. It appears, however, that as yet only one of the hotels has been kept open during the winter; and, as there have been hardly enough guests to make it profitable, it is doubtful whether any will be available for the coming winter.*

'The remaining village in the upper Engadine is Samaden, which stands at the height of 5600 feet above the sea, and about 100 feet above the valley of the Inn. It is well sheltered by the grassy slopes of high mountains to the north, and less completely to the east, and has the advantage of a fine view of the Bernina range to the west, which, being distant, does not shut out the sun so much as the nearer mountains to the south of St. Moritz. Still there is

* The Engadin-Kulm Hotel was kept open during the winter 1869-70, and several inmates remained with considerable advantage and enjoyment, beguiling their Arctic season with systematic arrangements for sledging and skating.

a high range due south of Samaden, and this is the common disadvantage of the whole Engadine during the winter, that high mountains to the south shut out much of the sun's light and heat. Another objection to Samaden is the somewhat marshy character of the valley at this part, the waters of the Inn sometimes spreading over its flat alluvium. I have seen, in the early morning, a thin stratum of fog brooding over this, like that on the lake and flat about the Kurhaus of St. Moritz. Still the Bernina Hotel stands sufficiently high above this not to suffer materially from its influence. I have already spoken favourably of this hotel and its manager, M. Franconi; and, as it is the only hotel which is constantly kept open during the winter, it must be considered the head-quarters of those who intend to make a long sojourn in the upper Engadine. I found in the *livre des voyageurs* several testimonials from those who had wintered there, speaking in strong terms of the benefit which they had gained in their health, and of their entire satisfaction with the care and attention bestowed on them by M. and Mme. Franconi.

'I have already adduced many proofs that the upper Engadine is too high and too cold even in the summer for the comfort and well-being of delicate invalids, and even of some persons in health; and it becomes a question whether sufficiently cool and bracing summer resorts may not be found at a somewhat lower elevation, and so accessible, and with such comfortable accommodation, as to make them available.

'Lower down in the valley of the Engadine, at an elevation of 4000 feet, is the large newly-built bathing establishment of Tarasp-Schuls, with accommodation for three hundred persons. The mineral waters, both for drinking and for baths, saline, chalybeate, and sulphurous, are the great attraction; the saline, which are also alkaline, being stronger than those of Carlsbad or Kissingen. But the situation may also be recommended as a cool summer residence, being shaded by lofty and well-wooded mountains to the south, with varied walks and drives on a dry soil, and with pure air. The high southern screen, however, unfits it for a winter abode, and it has no protection from the eastern blasts which come up the valley.

'There is a considerable hotel, or Kurhaus, at Davos am Platz, situated in a high valley to the north of the Engadine, near the Strela pass. It stands about 5000 feet above the sea, and is much frequented, on account of its cool pure air, and the whey-cure which is carried on there. I have passed some days there, and found the air pure and invigorating, where not contaminated by the effluvia of *liquid manure*, so profusely laid on the pastures in many parts of Switzerland—no doubt very good for the vegetation, but wholly spoiling the fragrance of the mountain breeze. How the composts or cesspools for these manures haunt you wherever you go among the high Alps! Even at Zermatt, that village of glaciers, at Breuil, on the other side of the Matterhorn—places from six to seven thousand feet high—and at most of the high mountain *chalets*, where you might expect to be above such nuisances, and to breathe the pure air of heaven, you find it poisoned by the stench from these cloacæ, in which purity and health are sacrificed on the altar of utilisation! However, although I happened to encounter this nuisance at Davos, it might have been accidental. The place has a repute for great purity of air. One of my phthisical patients spent six weeks there this summer, and improved in appetite and general health, but without change for the better in the lungs; and he complained much of the dulness of the place. It does not appear to be suitable for winter residence, as it has little or no shelter from cold winds.*

---

* Dr. Mayer-Ahrens gives the mean temperature of the whole year at Davos at 37° Fahr. Nevertheless, Dr. Spengler, a physician resident at Davos, states that nearly a hundred guests have passed the last winter there 'with gratifying results.' ('Med. Times and Gazette,' April 9, 1870, p. 405.) Dr. S. also asserts with confidence that several patients with pulmonary disease had been greatly benefited, and some cured during their stay at Davos.

'It was a chief object of my present tour to find a high site for summer sojourn, less in extremes than the Engadine, better sheltered from the bleak north and eastern winds; and the new baths of Bormio, on the south ascent of the Stelvio pass, seemed likely to answer to this description. The pamphlet of my friend Dr. Fedeli ('Sulle Acque Termali e Fanghi di Bormio'), and the fuller treatise of Drs. Meyer-Ahrens and Brügger ('Die Thermen von Bormio'), both published in the present year, had directed my attention to this place; and to these I must refer for details respecting the composition and properties of the waters, and their sundry applications for the cure of disease. My concern was chiefly with the situation as a mountain residence for the many in summer, and possibly for the few in winter also.

'I may as well take the reader with us in our way from the Engadine. On August 23, we left Samaden at five o'clock, to cross the Bernina, on a fine morning, with not a cloud to be seen; but the valley of the Inn and the grass-fields of Pontresina were white with frost. And, mounted on the *banquette* of the diligence, and whilst gazing with admiration on the magnificent views of mountain-peaks, pine-woods, snow-fields, and glaciers, which the zigzags command in such variety, I was right glad to rise above the cold shade of the Piz Languard group into the sunshine of the little valley in which lie, side by side, the Lago Nero and Lago Bianco—the one pouring its waters into the Adriatic, the other into the Black Sea. Above these, near the summit of the pass, at a height of 7600 feet, stands the new Bernina Hotel, with fair accommodation, to tempt those aspiring individuals who do not find the Engadine high enough. It is a bleak dreary spot, more suitable to Alpine Club men than to invalids.

'Beautiful as are the views in the ascent of the Bernina pass from Pontresina, those on the descent to Poschiavo are even more striking, from the greater depth and steepness of the valley into which the zigzag and serpentine versatility of the road carried us. At first, from the giddy heights of savage mountain-tops, diversified only with snow-peaks and glaciers, we gaze down into the blue haze beneath, and, in the far distance, discern the tracery of a rich valley, speckled here and there with buildings and towers, looking like a fairy-land below us. Not many minutes elapse, as we are rapidly whirled down, before we find ourselves among trees and shrubs—quite novelties, after the scanty silva of the Engadine. Birch, beech, ash, elm, oak, and chestnut, now appeared in succession; then acacias, contrasting their varied hues of verdure with the sombre pine and dark brown rocks; and soon the vine, the gourd, and other luxuriant creepers, gave tokens of the entire change of climate which we had made in this rapid descent. Poschiavo, the first town in the valley, is only 3300 feet above the sea; so that, in little more than an hour, we had made a descent of more than 4000 feet.

'Poschiavo is not a good halting-place; but three miles beyond, on the brink of the lovely little lake of the Poschiavino, stands the very comfortable hotel and bathing establishment of Le Prese, in the manager of which I was pleased to recognise an old acquaintance, M. Mella, formerly the proprietor of the Grande-Bretagne at Bellagio, on Lake Como, who had made that delightful hotel so attractive to travellers. The mineral water is alkaline, sulphurous, and very slightly chalybeate. It is a cold spring; but there is a steam apparatus for heating the water for the baths, which are newly constructed of Italian marble. The waters are said to be particularly efficacious in dyspepsia, skin-diseases, chronic rheumatism, and scrofula.

'The moderate height of Le Prese—about 3000 feet—and its situation in a narrow valley flanked on either side with lofty mountains, would prevent us from considering it a very bracing or invigorating place; but it is much cooler than the Italian lakes generally, and presents a pleasant intermedium for spring and early summer, between those and the higher stations, and may prove cool enough in summer for many who have passed their winter in the Riviera or in South Italy.

'From Le Prese the road descends 1500 feet in ten miles to the opening of the Val di Poschiavino into the Valtelline at Tirano, a thoroughly Italian town. From this the road gradually ascends all the way to Bormio; and the sooner travellers can get up the Valtelline, the better; for, like all the low Italian valleys, it is as malarious and unhealthy to animal life as it is luxuriant and rich in vegetation. The tumultuous Adda, fed by a thousand mountain torrents, is continually depositing an alluvium of mud on all the flatter parts of its course, which, forming a rich swampy soil, becomes, in the summer-heat, a hot-bed for vegetable growth, but a reeking source of miasmata to man. It is, indeed, a sad and striking contrast to see under luxuriant vines, hanging in rich clusters among the gigantic maize-stalks, all teeming with fertility, the sallow peasants, with haggard faces and goitrous necks, many stunted and deformed, and very 'few presenting the complexion or configuration of health. This remark applies more to the lower Valtelline, from Tirano to Colico, than to the upper portion; and the improvement in the aspect of the inhabitants as we ascended the valley was very remarkable. At Bolladore, the flat part of the valley, with its exuberant fertility, ceases; and, after passing through a fine grove of Spanish chestnuts, the road enters the narrow defile of La Serra, with its stupendous walls of perpendicular rock, separating the rich Valtelline from the *paese freddo*, as the upper or Bormio end of this valley is termed. This, athough bounded on every side by lofty mountains or savage rocks, is not wanting in brightness or fertility. The valley, emerging from the gorge, is first narrow, with wooded knolls beneath the towering mountains, but above expands into an undulating plateau of green fields, beyond which, on the right front, stands the dilapidated little city of Bormio; and to the left, on a rocky slope, the large establishment of the Bagni Nuovi, with the irregular buildings of the Bagni

Vecchi beyond, at a greater height; niched at the foot of the great rocky barrier of the Stelvio mountains, which here terminate in perpendicular cliffs, as rugged and savage as weather-scarred limestone can make them.

'It must be confessed that this first view of the Baths of Bormio from below is too wild and desolate to be beautiful; but, as the Stelvio road emerges from the narrow streets of the ugly town, and ascends the gentle slope to the baths, the scene all brightens; and, when we reached the terraced gardens of the building, and, turning our backs on the frowning Stelvio, looked in front over the green fields to the valley and gorge that we had left, now in the azure distance, with its varied mountains towering above it, some wooded, some barren, and most of them capped with snow —then another beautiful valley, with its vista of mountains lying to the left; and withal, for a foreground, a garden glowing with grouped flowers of brilliant colours—we were constrained to admit that we had surveyed few scenes more lovely in the lands which we had left behind us. Thus were removed our first misgivings as to the beauty of the place.

'But next as to its salubrity. The establishment of the New Baths is a modern building, placed on a hill sloping up to the perpendicular cliffs before mentioned, which are the terminal buttresses of Monte Cristallo, the highest of the two mountains between which the Stelvio road is carried. This hill is formed of the *débris* of the limestone rock fallen from the cliffs, and is consequently dry and stony, with large masses of rock scattered here and there over it. A thin soil, with mountain herbage and scattered shrubs, partially covers it; and this has been further improved by artificial planting and a variety of walks and seats for the benefit of the inmates of the establishment. The building faces the south, looking down the Val di Sotto or Upper Valtelline. Behind, the rocky cliffs of the Stelvio form a complete screen to the north and east. To the right is the valley of Pedenos, running westwards, and leading to the Vals Viola and Livigno, through which it is possible to make short cuts to the Bernina road. To the left, beyond a projecting spur from the Stelvio mountain, lies the valley of Furva, which runs up to Santa Caterina, another watering-place, to be noticed hereafter.

'The new baths of Bormio are, therefore, favourably placed for dryness, sunny aspect, and shelter from the winds of the north and east; and yet derive coolness from the altitude, and from the breezes which frequently blow from the west and from the south. No doubt the heat of the sun is great in summer; and there is a want of tree-shade near the establishment. But the detached rocks afford shelter; and there are always portions of the Stelvio road in the shade, as it winds through the ravine behind the old baths. On the opposite hill, also, are extensive pine-woods, in which shady walks may be found.

'The meteorological observations collected by Dr. C. G. Brügger give a favourable view of the temperature of the Bormio baths. The mean annual temperature is 44·51 deg. F., which is from two to four degrees warmer than any place of the same altitude in Switzerland. The moderateness of the temperature at different seasons is further proved by these figures:—

'Temperature of the Air at Bormio (New Baths).

	Max.	Min.	Mean.
Summer	F. 80 deg.	41 deg.	61·7 deg.
Autumn	73·2	18·5	43·16
Winter	—	—	31·6
Spring	70	21	42

'The mean humidity of the air at Bormio in the three summer months is 68 deg. (saturation being 100 deg.). By way of comparison, may be mentioned that of Berne (75·7 deg.), and those of Zurich (79·8 deg.), and Montreux (80 deg.)

'These figures correspond pretty well with the indications of our sensations during our sojourn. The air was never either oppressive or too cold to be pleasant. The sun's rays, of course, had great power; but there was almost always a cooling breeze, which was refreshing, without the extremes of scorch and chill from which we had suffered in the Engadine.

'The airiness and dryness of the situation may be ascribed partly to its absolute height (which I found 4560 feet above the sea; Ball states, 4798), and partly to the declivity falling from it—on the west, abruptly into the ravine down which dashes the torrent of the Adda; and on the south, by a more undulating slope, into the Bormio valley; while behind, to the north and east, it is protected from the colder winds by the Stelvio mountains. It may be said to stand in an amphitheatre of mountains; but those in front are more distant, so that they shut out the morning and evening sun less than in the Engadine; whilst those behind are so near and so high as effectually to shut out the cold blast from the north. Hence the thermometer, in the coldest months, falls only a few degrees below freezing; whereas at St. Moritz it sometimes is down to -18 deg. F., or fifty degrees below freezing. The occurrence of frost at night during the summer months, which we had experienced in the Engadine, is never thought of at Bormio. In fact, the growth of gourds, and of several other tender plants, in the open gardens of the establishment, proves how much more genial is the climate than that of the upper Engadine, where even potatoes and common garden-stuff cannot be raised.

'On comparing Bormio with other places, with regard to the amount of rain and the number of rainy days in the year, it appears from Dr. Brügger's tables to be much below the average; the number of rainy days in the summer (June, July, August), during five years, averaging twenty-three; whilst at Zurich it was thirty-two; at Berne, forty-five; Zermatt, thirty-eight; Remüs (Unter Engadin), forty-two; Gastein (Tyrol), forty-three; Tegernsee (Bavaria), fifty-two. In predominance of fine weather, therefore, Bormio resembles Italy more than Switzerland. Possibly the paucity of trees in the immediate neighbourhood may be a cause of this greater exemption from rain and cloud. I was struck with the difference between this place and Switzerland generally, in the prevalent prognostic as to the weather. In Switzerland, it is commonly doubtful or gloomy. At Bormio, in spite of occasional gathering clouds and falling rain, the general assurance was, 'It will soon be fine again;' and so it proved.

'The establishment of the Bagni Nuovi is under the management of an intelligent Swiss, who speaks English, and is a most attentive host. It contains one hundred and forty bedrooms, plainly but comfortably furnished; and there are the *salles à manger, salon de société, salle de lecture*, billiard-room, &c., usually found in large continental hotels. There are forty bathing-rooms, with baths, some of marble, some of wood; and appliances for douches of different kinds; and a few are appropriated to collect the muddy deposit from the waters (*fanghi*)—a disgusting-looking slimy matter, redolent of sulphuretted hydrogen, supposed to be effectual as a discutient for tumours and rheumatic swellings.

'It is foreign to my purpose to describe the waters and their uses, and I must refer to the pamphlet of Dr. G. Fedeli, the intelligent physician residing in the establishment; and to that of Drs. Meyer-Ahrens and Brügger, before mentioned. I may merely state that the thermal waters gush in great abundance from several sources in the tufa deposits at the foot of the Stelvio mountain. These may be explored half a mile above the new baths, at the Bagni Vecchi, a smaller and still more economical establishment. The temperature of the springs here is as high as 106 deg. F. At the Bagni Nuovi, to which it is conveyed in pipes, I did not find it higher than 98 deg. F. The chief mineral ingredients in the waters are sulphates of lime, magnesia, and soda, with very little iron and sulphur. The mud, however, which is deposited from the

Archduchess Spring, is rich in filaments of sulphur, and in sulphuretted hydrogen, besides ochreous and saline matters. There is also a very drinkable cold chalybeate spring just below the old baths, not unlike that of St. Moritz, but less brisk with carbonic acid. It is much recommended for the weaker patients who are using the baths.

'The establishment of the new baths is open from the middle of June to the end of September; but, if the advantages of the situation for purity and dryness of air—its sunny aspect, yet airiness—its coolness, without bleakness—were more known, its season would probably be lengthened, and part of the establishment kept open through the winter.*

'There are many delightful excursions in the neighbourhood deserving of notice; but the limits of this communication will allow me to mention only that to Santa Caterina, another hotel or health-establishment, with accommodation for fifty inmates. There are only two or three baths; but the attraction is a noted chalybeate spring. It seemed to me too strong to be taken without dilution; but it is bottled to a great extent, and is much used as an addition to wine or to other waters. The distance is about ten miles from Bormio, up the beautifully-wooded glen, Val Furva. Although St. Caterina stands at a height of 6000 feet—as high as St. Moritz—there is in this glen leading to it a much greater variety of trees than in the upper Engadine. But, as at St. Moritz, the mineral spring rises out of a bog; and, as you walk to it, the path yields with singular elasticity under your feet. This swampy ground forms the bottom of the valley, which is closely hemmed in by lofty mountains on all sides, except to the west, where it leads down the Val Furva towards Bormio; and to the northeast, where it is open to the ice-bound Val Forno, sur-

---

* In the present year (1870) the establishment was opened at the end of April, and it will probably be kept partially open during the next winter.

rounded with the snow-peaks and glaciers of Monte Cevedale, Zufall-Spitz, Monte Trescro, and others. However advantageous this situation may be for scenery and mountain-climbing, it is obviously not a suitable residence for invalids, or for any persons likely to suffer from cold and damp.

'Besides the route which we took to Bormio over the Bernina pass, there is another down the Engadine to Martinsbruck, and by the Etschthal over the Stelvio pass, the highest and grandest of all the carriage-passes over the Alps. Or the Stelvio may be reached from Innsbruck by the Brenner railway to Botzen, and up by Meran and the lower part of the picturesque Etschthal. From the Italian side it is possible to go from Varenna or Bellagio on the Lake Como to Bormio, in a long summer's day; but the journey is generally divided by sleeping at Sondrio, which is the least unhealthy of the towns in the beautiful but dangerous lower Valtelline, in which it is not expedient to linger even for a day.

'Although the new baths of Bormio appear to me to offer more advantages in point of dryness, shelter, and comfort than any of the other high mountain resorts in the Alps, yet several of these deserve mention, as affording good accommodation, and being entitled at least to compete with the Engadine as summer quarters.

'The Hotel Rigi Kaltbad (4727 feet) is on the south-west of the Rigi, and thus sheltered from the coldest winds. The Hotel Rigi Scheideck (5406) is on the south-east of the mountain, less sheltered, but is quiet, with more scope for promenade, and well supplied with milk and whey. The great objections to the Rigi hotels are, the frequent occurrence of bad weather, and their inaccessibility by carriage.

'Leukerbad, under the Gemmi pass, stands at a height of 4642 feet, with a southern aspect, and by its great mountain-screen to the north and east, is fairly protected from extreme cold. It is accessible by a good carriage-road from the Rhone Valley, and has good hotel accommodation.

'Courmayeur, on the Italian side of Mont Blanc, stands at a height of about 4000 feet, with two large hotels, but it is too close to the Brenva glacier and the great snow-fields of the Géant to be safe from sudden chills.

'At Gressoney St. John, in the Val de Lys, is Delapierre's very comfortable hotel, at a height of nearly 5000 feet, in a beautiful Alpine valley, with the Lyskamm at its upper end; but this splendid snow-mountain, which is its pride, might send down its bleak blast at times, to the detriment of the delicate. The valley at present is not accessible to carriages.

'There is a good hotel on Monte Generoso, between the lakes of Como and Lugano, which may prove a good summer residence. The summit is 5561 feet. This also is accessible only on foot or on horseback.'

Comballaz, already mentioned as an accessible and comfortable station in the Diableret district, requires notice here also, as its height, 4416 feet, entitles it to rank among the highest class.

Lauenen, at a height of 4134 feet, in the upper valley of the Sarina, can be reached from either Thun or Vevey, and lies in an amphitheatre of high peaks near the Gelten glacier, with a clean rustic inn.

I have just visited Evolena (4700) in the Val d'Herens, out of the Rhone valley, near Sion, and found fair accommodation at the Hôtel de la Dent Blanche, with fine views of the snow mountains. A char-road to it will be open in August. Although a delightful change from the Rhone valley, the heat was rather overpowering even here; and I may remark that the long days and greater height of the sun in the present month (June) render these places amid high mountains hotter than they are in the month of August, when, although from accumulation heat becomes greater, yet the heating operation of the sun is of shorter duration.

Andermatt (4642) and Hospenthal (4787) on the St.

Gothard road, deserve mention as of the requisite height and of easy access; but they are too much occupied by passing travellers to be suited for a sojourn for invalids. The same remark applies to the inns of the Rhone glacier (5465), the Furca (7911), the Splugen (4757), and the Bernina (7600).

There are several well-known mountain inns accessible only by mule-path, and very attractive for the grandeur of their scenery, but not suitable for most invalids Such are the Æggischhorn (7000), the Belalp, the Riffel (8428) of Zermatt, the Rigikulm (5905), Pilatus (6282), and Mürren (5347). The more vigorous of valetudinarians may be able to accomplish a visit to them by way of excursion; but for a sojourn they would prove too uncomfortable, and in bad weather dangerous.

The Pyrenees have very few habitable abodes at a great height; Barèges and Çauterets only reaching to from 3000 to 4000 feet. Gavarnie and Panticosa rise to 4000 and 5000, but are wanting in accommodation suitable for pulmonary invalids.

I have thus collected, in a somewhat desultory manner, notes derived from my own observation and that of others, which may, I trust, prove useful to those who seek to improve their health by the aid of climate. The original and chief purpose of this little work was to give a perfectly *impartial* view of the comparative qualities of the principal places in the south of Europe in which the *winter* is so much milder and drier than in England, that invalids, especially pulmonary invalids, may there pass through that trying season, without that deterioration which pretty surely awaits them at home. And with respect to the utility of this measure, which has been sanctioned by the practice of centuries, I have found in the records of a large experience to which I have access, abundant evidence of satisfactory results. On the other hand, Dr. H. Weber quotes the opposite opinion of an eminent London physician to this effect:—'My experience with regard to the warmer health resorts is great; but it is unfortunately not favourable.' In answer to this, I must refer my readers again to my first five chapters, and here merely add, that in so formidable a disease as phthisis, too much must not be expected from *climate alone*. *All the aids* to be derived from *medicine, diet,* and *regimen* are required; and when these are fully and fairly used, the results have been far from unfavourable in a large majority of cases. But when further good results are reported by Dr. H. Weber and others, as accruing likewise *from the high mountain treatment of consumptive cases in winter as well as in summer*, in common with all interested in the subject, I am most desirous that this plan should be fairly and favourably tried; and to supply information which may aid in fulfilling this object, I have added this Appendix; indicating the several places in Switzerland and North Italy which offer the best summer quarters for invalids, and some of which may be available in winter also.

78, *Park Street, Grosvenor Square, July* 1870.

www.ingramcontent.com/pod-product-compliance
Lightning Source LLC
Chambersburg PA
CBHW032155160426
43197CB00008B/924